Sketches

of Western Canada

1873-1899

Geology & Landscapes

George Mercer Dawson

LL.D. CMG FRS FRSC

Peter D. Geldart, editor

Sketches of Western Canada, 1873-1899.
George Mercer Dawson, LL.D. CMG FRS FRSC
Collected and prepared by Peter D. Geldart
geldartp@gmail.com

Petra Books
78 George Street, Suite 204
Ottawa ON K1N 5W1 Canada

Front cover: Geological Survey of Canada field notebooks 1884, G. M. Dawson. National Archives.
L: "The front of glacier on (Stickeen) [sic] river about 3 mi long according to McConnell."
R: "Upstream end of Great Glacier", British Columbia.

Back cover: Personal notebooks, G. M. Dawson, McGill University Archives, Dawson-Harrington fonds MG1022-6-81, p 35, August 16, 1874. Mt. Yarrell from the forks, British Columbia.

Introduction

The sketches from George Mercer Dawson's (1849-1901) notebooks in this collection are largely on the subject of geology and the survey of land and waters, plus a portion on indigenous peoples, in which he had a special interest. His diaries, letters and personal notebooks were originally scanned by McGill University Archives for the Dawson-Harrington fonds. Geological field notebooks are at the National Archives. The period covered begins after graduation from McGill University followed by the British School of Mines. He turned down an appointment (aided by his father) to the Geological Survey of Canada (GSC), which he later took up, in preference for outdoor scientific work with the British North America boundary commission. He was at the GSC as a geologist from 1875 to his premature death in 1901, and served as Assistant Director from 1883 and Director from 1895. He took every opportunity, largely except in winter, to travel west to survey the land. He was supported by a team with all necessary supplies. This took him to British Columbia, the northwest coast, the Yukon, Alaska and beyond.

Key dates:

British North America boundary commission. 1873-1874
https://www.ulethbridge.ca/lib/digitized_collections/ourheritage/index_page_stuff/Following_Trails/Dawson/Dawson_74/Dawson_1874_1.html
Joined the Geological Survey of Canada (GSC) 1875
Appointed Assistant Director of the GSC 1883
Queen Charlotte Islands survey. 1878
Appointed Director of the GSC 1895
Behring [sic] Sea Commission. 1891-1892 with co-scientists Baden-Powell and Dawson
https://open.library.ubc.ca/collections/bcbooks/items/1.0222471

George Mercer Dawson

Raised in upper-class Victorian Montreal (father Sir William was principal of McGill University) he was well educated and, beyond his career as a scientist, was good at drawing and painting, better, perhaps, than many other scientists working in the field. It was his task to record and assess what he saw and he allotted himself the time to do this. His field notebooks are quite detailed, in stark contrast to a notebook during a sojourn to France which is perfunctory: "Monday. Preliminary drive over Paris. Lunch. Walk in Boulevards. Gardens of Tuileries & Place Vendome. Tuesday. Spent all day in the Louvre. Call at H. Strode around streets to Palais Royal." (McGill 1022-6-136 1880 p 102). He seems to have preferred exploring the outdoors wilderness.

In this review of the available notebooks at McGill and the National Archives (Dawson never published such a collection) some sketches which were quite similar to others or which were low-resolution or of little interest graphically have not been included. Note that the National Archives states that some field notebooks are missing: "A concerted effort was made by the Geological Survey to locate these volumes through contacting former geologists as well as those currently surveying in the same area."

Not included in this collection: Dawson also had prepared a few dozen Dallastypes (a form of etching) of his drawings intended for use in publications. Please see https://archivalcollections.library.mcgill.ca/index.php/drawings-for-dallas-types-undated. Nor are his watercolours found in the notebooks included here.

Categories have some overlap and are not intended to be precise. The images are ordered by date, with some concessions to layout. The actual locations of the sketches may be able to be determined by reading the notes on the image, if any, or the adjacent script in the notebooks. Closer examination of the text and images from the McGill University Archives can be made online at https://archivalcollections.library.mcgill.ca/index.php/george-mercer-dawson-papers referencing 1022 (Dawson-Harrington) -6 (George Dawson) -pXX (page number in the PDF). The GSC field notebooks can be consulted on site at the National Archives (Ottawa). Images taken with a camera on site or captured online have been slightly enhanced and sharpened for legibility.

Legend in the captions:

"McGill University 1022-6" = McGill University Archives, Dawson-Harrington fonds, online archives. personal diaries and letters. Fonds MG1022, George Mercer Dawson -6, File #xxx, page #. https://archivalcollections.library.mcgill.ca/index.php/george-mercer-dawson-papers

"Geological Survey of Canada field notebooks" = National Archives of Canada RG45 (George Mercer Dawson) Geological Survey of Canada.
https://recherche-collection-search.bac-lac.gc.ca/eng/home/record?app=fonandcol&IdNumber=136921

— PG

Contents

These general categories allow the presentation of the sketches, but are not intended to be precise. The images are arranged by ascending date, with some concessions to layout.

George Mercer Dawson

Landscapes

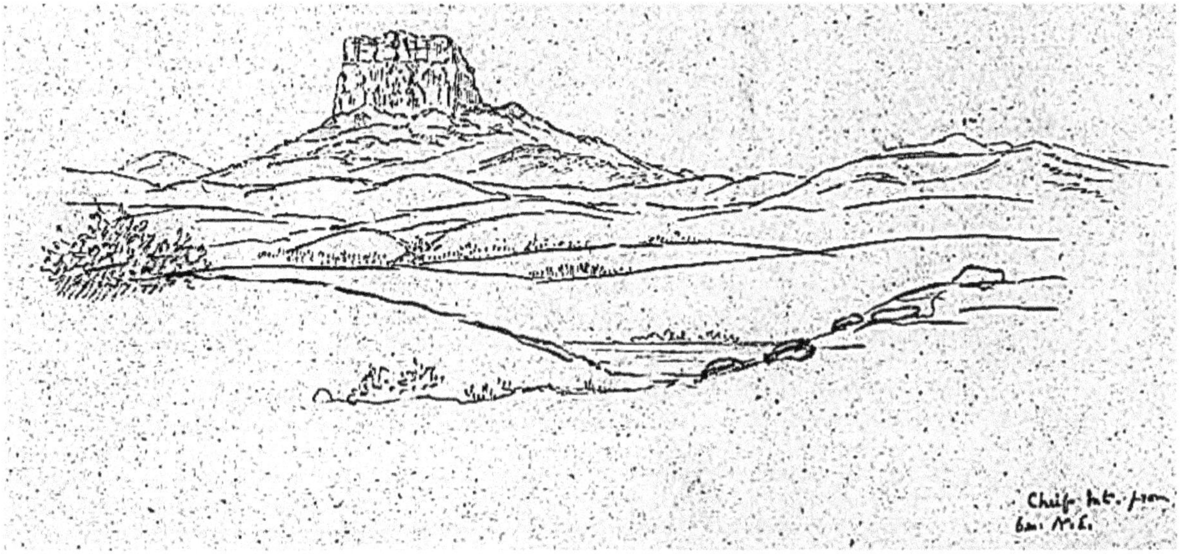

Dawson 1874. McGill University 1022-6-81 p52

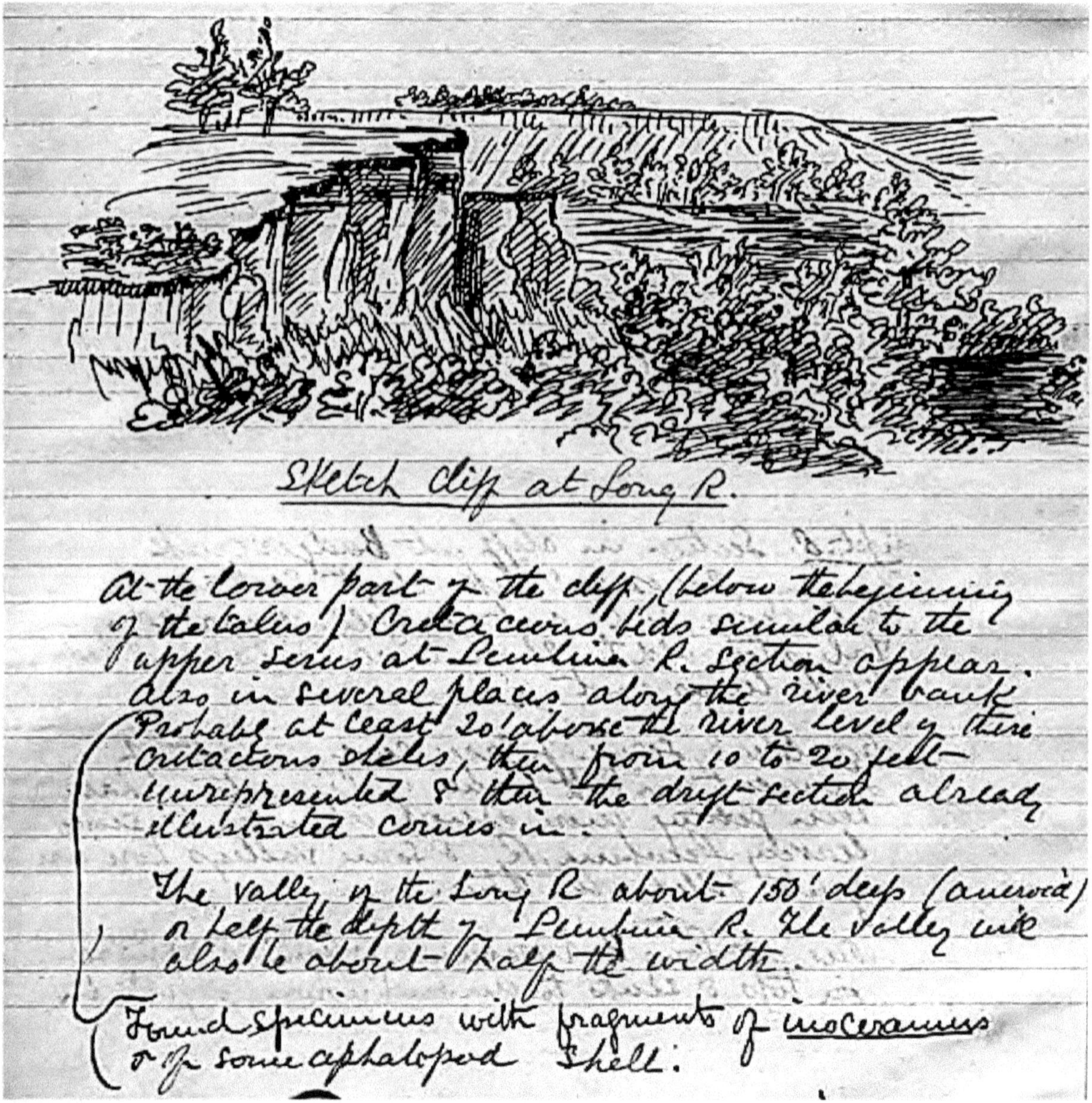

Sketch cliff at Long R.

At the lower part of the cliff (below the beginning of the talus) Cretaceous beds similar to the upper series at Pembina R. Section appear. Also in several places along the river bank. Probably at least 20' above the river level of these Cretaceous shales, then from 10 to 20 feet unrepresented & then the drift section already illustrated comes in.

The valley of the Long R about 150' deep (aneroid) or half the depth of Pembina R. The valley will also be about half the width.

Found specimens with fragments of Inoceramus or of some cephalopod shell.

Dawson 1873. McGill University 1022-6-177 p288

Thought at first that Cart tracks but soon found that could not be, but really old buffalo paths & probably indicating pretty well the direction of their migrations. The ruts were nearly parallel, generally ran along the low ground & all tended to the N.W.

Wednesday Sept 10. Got carts & harness put in repair, & stores replenished. Decided to leave Duckworth at the depot to shoot & skin & collect whatever he can. Thinking this better than to keep him travelling on the chance of reaching a better place

Dawson 1873. McGill University 1022-6-177 p19

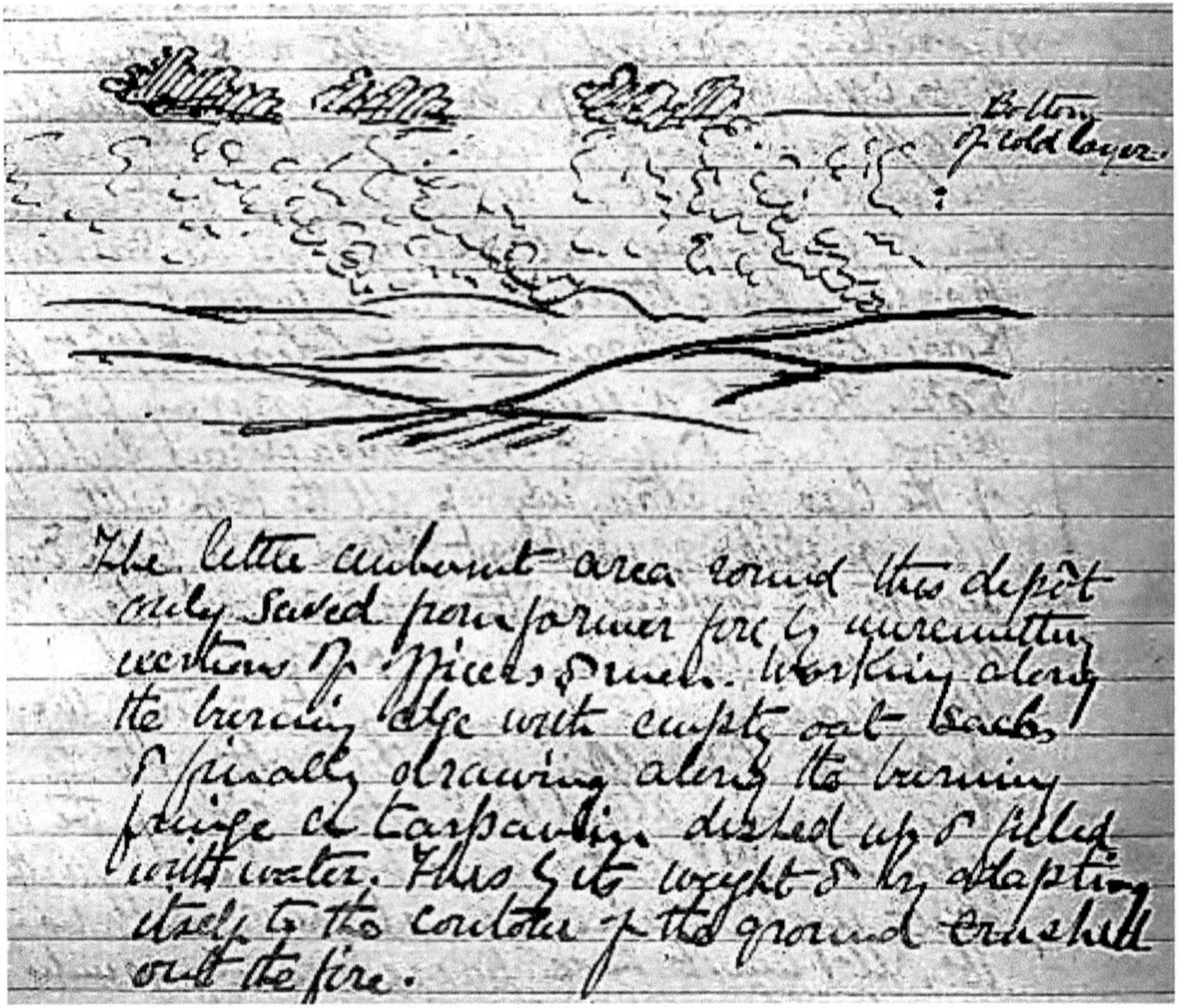

The little unburnt area round this depot only saved from former fire by unremitting exertions of officers & men. Working along the burning edge with empty oat sacks & finally drawing along the burning fringe a tarpaulin dusted up & filled with water. This by its weight & by adapting itself to the contour of the ground crushed out the fire.

Dawson 1873. McGill University 1022-6-177 p56

Dawson 1874. McGill University 1022-6-180 p91

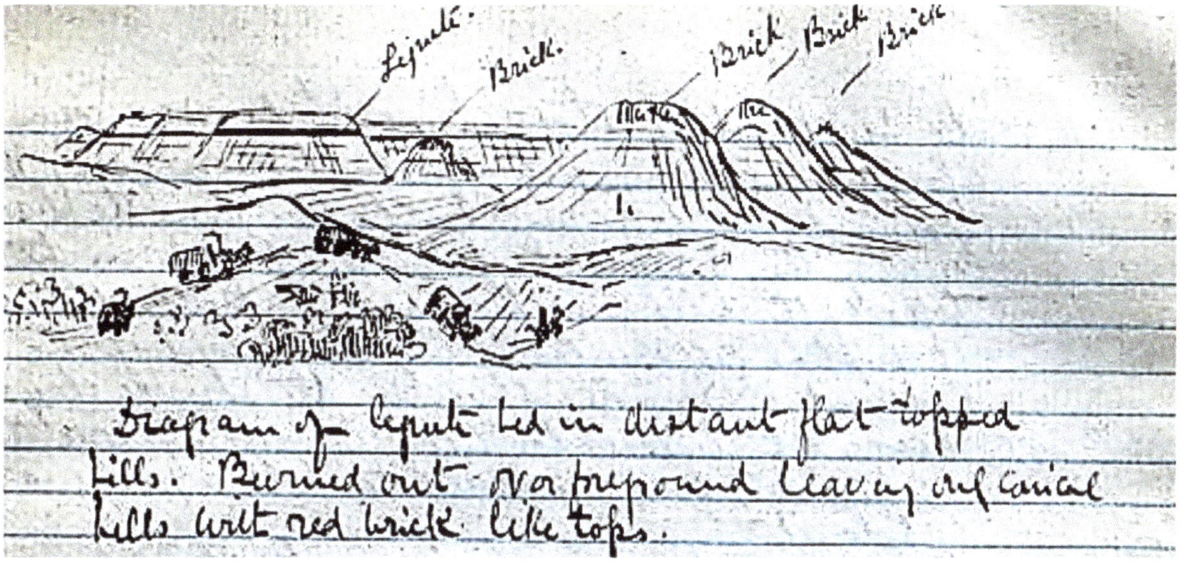

Dawson 1874. McGill University 1022-6-180 p59

Dawson 1874. Souris River, Sask.
McGill University 1022-6-81 p9

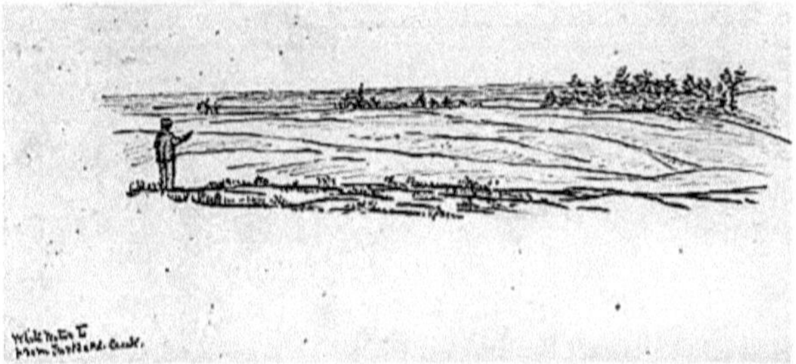

Dawson 1874. McGill University 1022-6-81 p7

Dawson 1874. McGill University 1022-6-180 p69

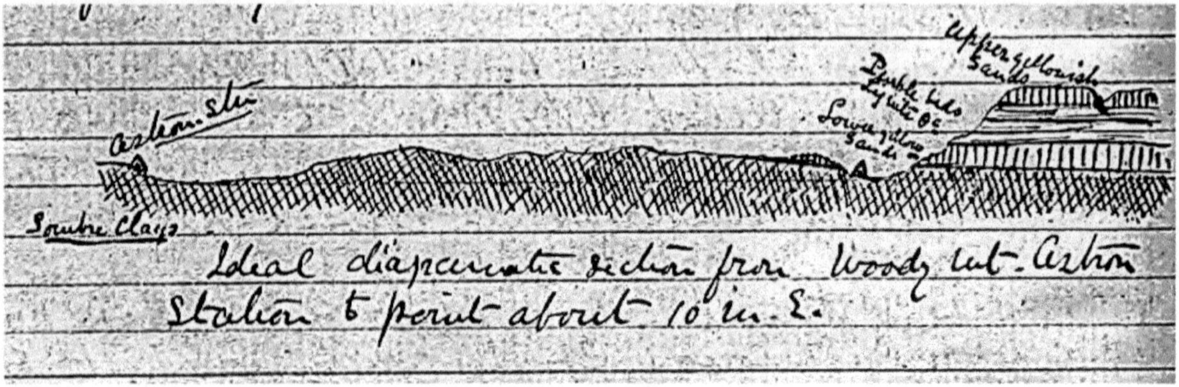

Dawson 1874. McGill University 1022-6-180 p60

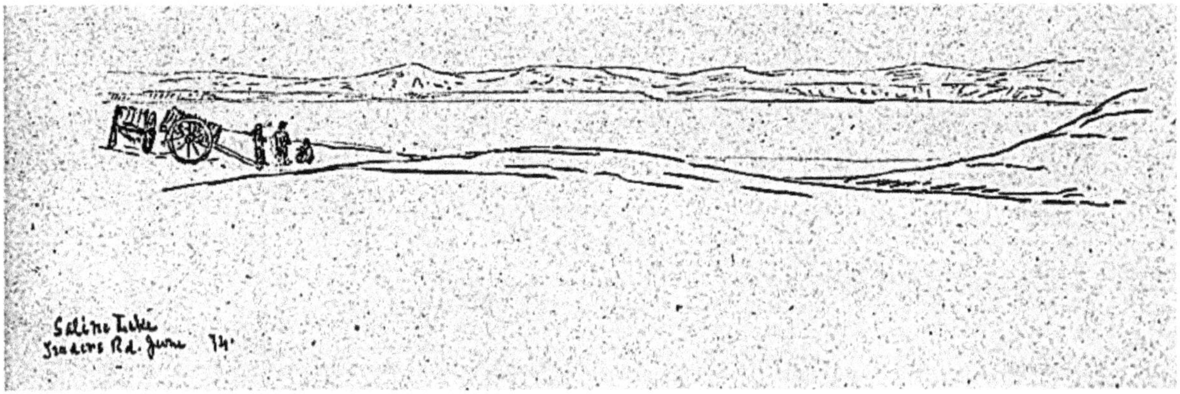

Dawson 1874. McGill University 1022-6-81 p12

Dawson 1874. McGill University 1022-6-81 p19

Dawson 1874. McGill University 1022-6-81 p49

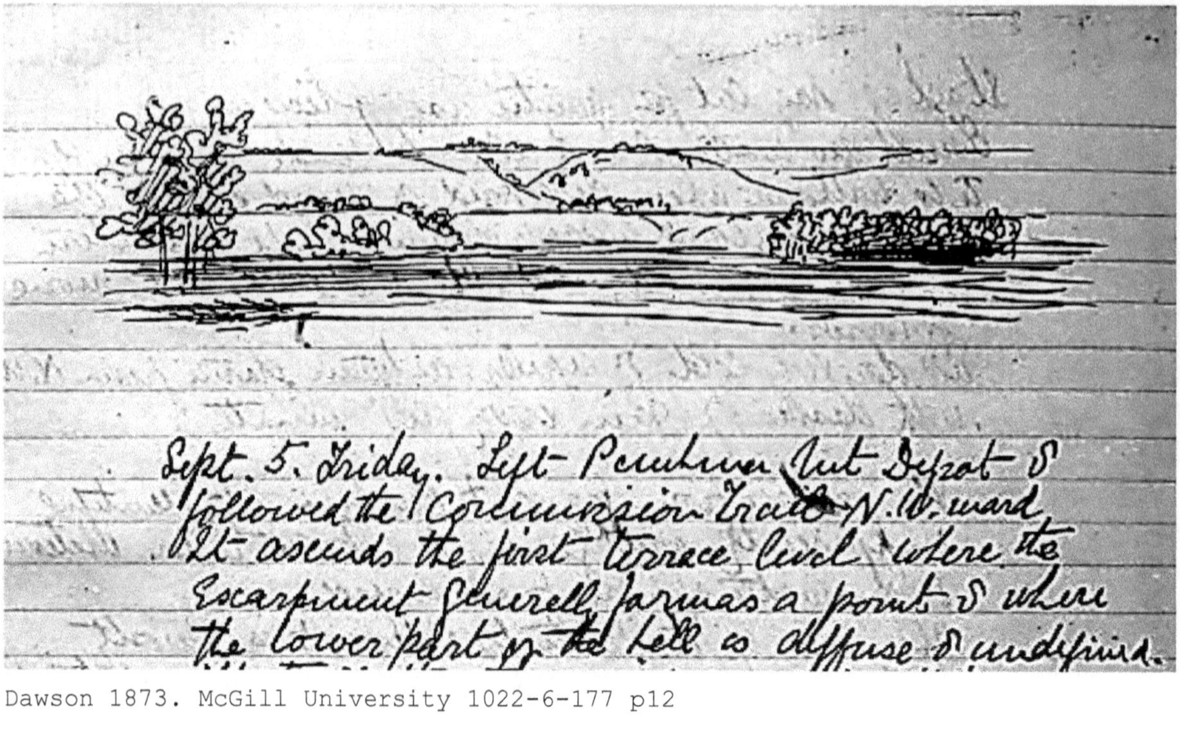

Dawson 1873. McGill University 1022-6-177 p12

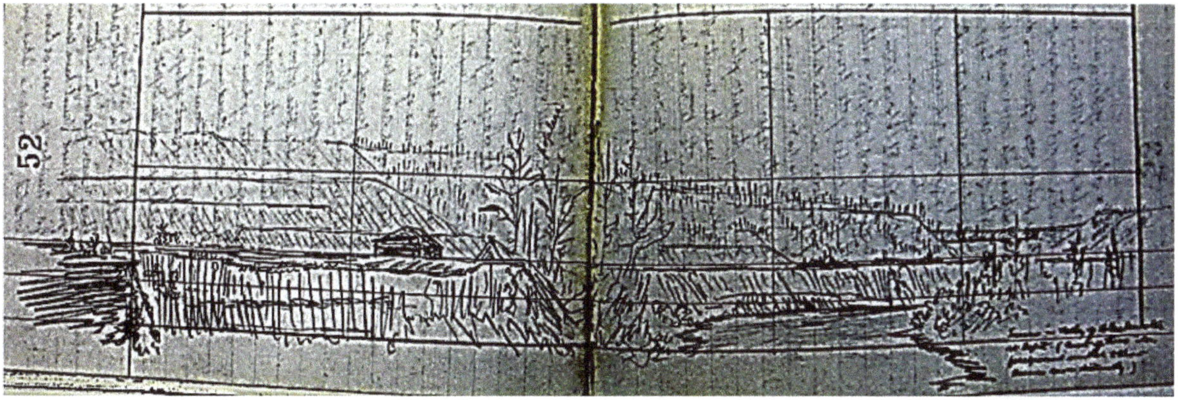

Dawson 1875. Blackwater Valley terraces. Geological Survey of Canada field notebooks.

Dawson 1875. McGill University MG1022-6-181 p59

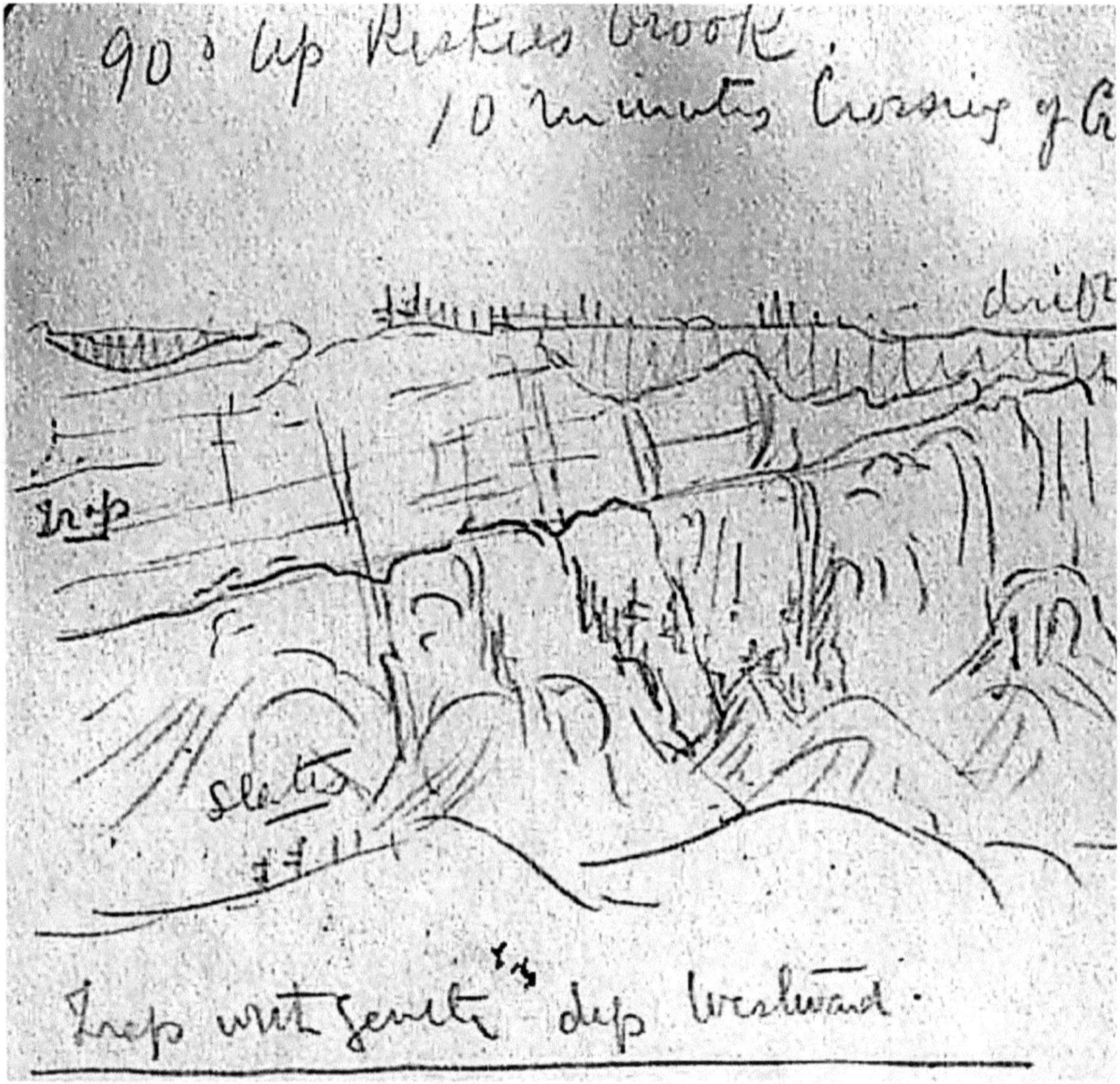

Dawson 1875. McGill University MG1022-6-181 p54

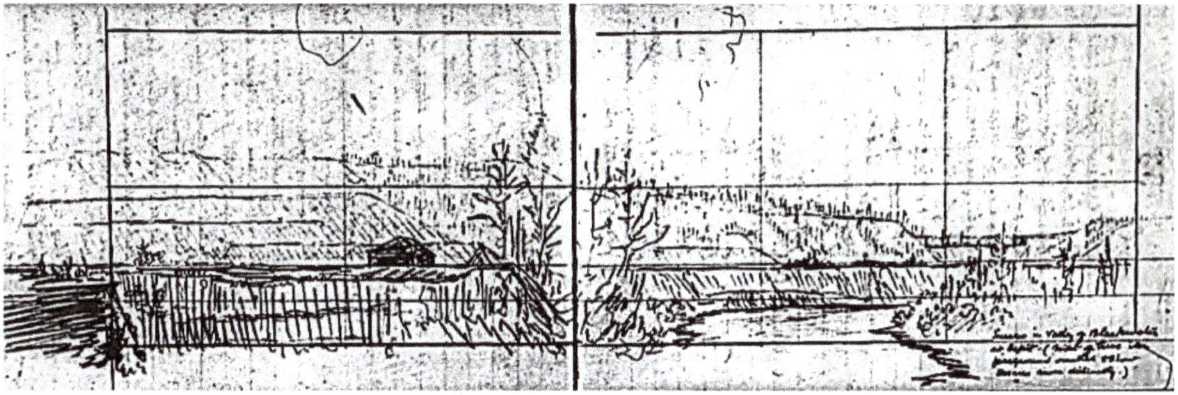

Dawson 1875. Geological Survey of Canada field notebooks.

Dawson 1875. McGill University MG1022-6-181 p94

Dawson 1875. McGill University MG1022-6-181 p78 LEFT

Dawson 1875. McGill University MG1022-6-181 p79 RIGHT

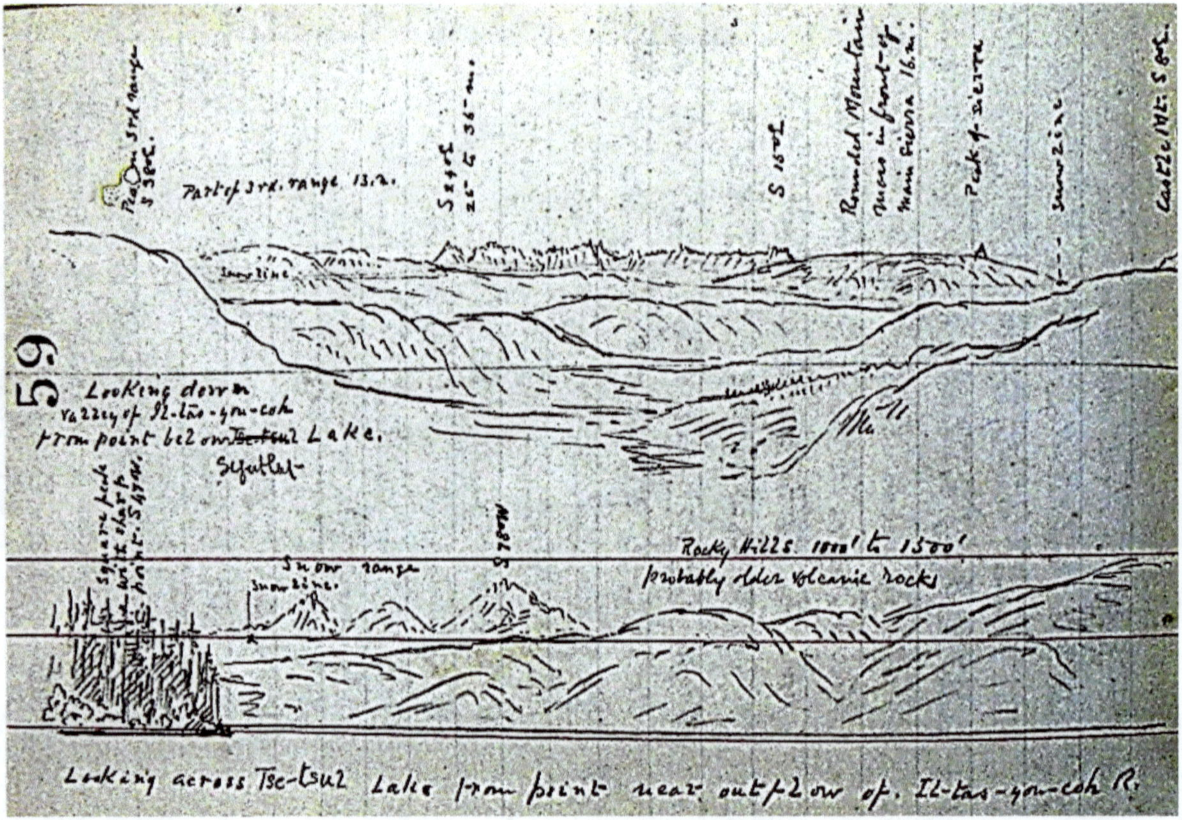

Dawson 1875. Vancouver Island. Tse-Tsuz Lake. Geological Survey of Canada field notebooks.

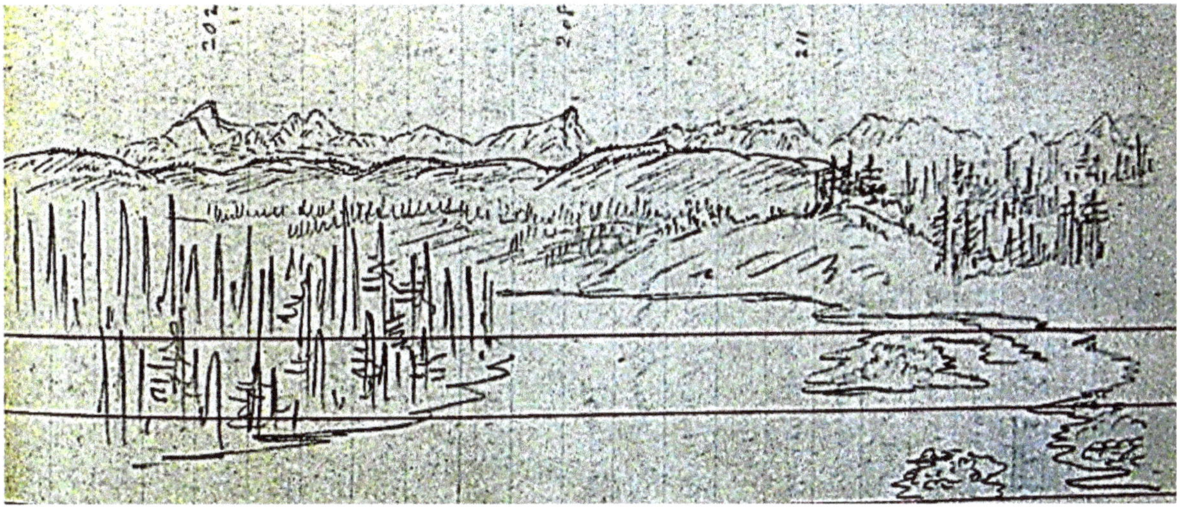

Dawson 1875. Vancouver Island. Tse-Tsuz Lake. Geological Survey of Canada field notebooks.

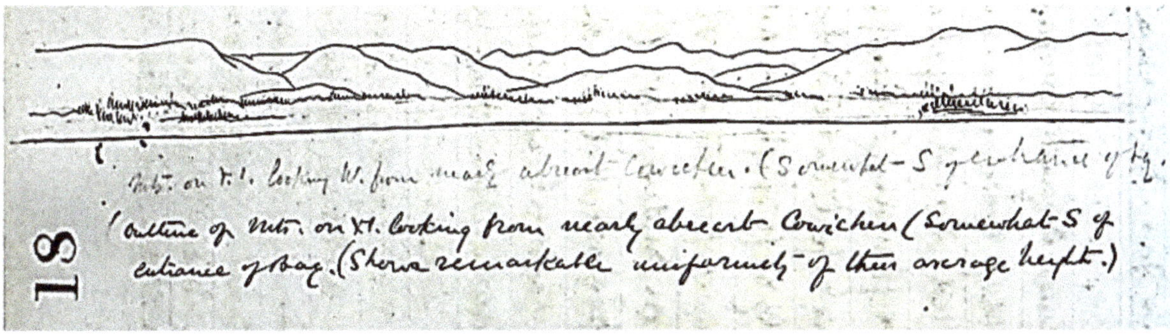

Dawson 1875. Geological Survey of Canada field notebooks.

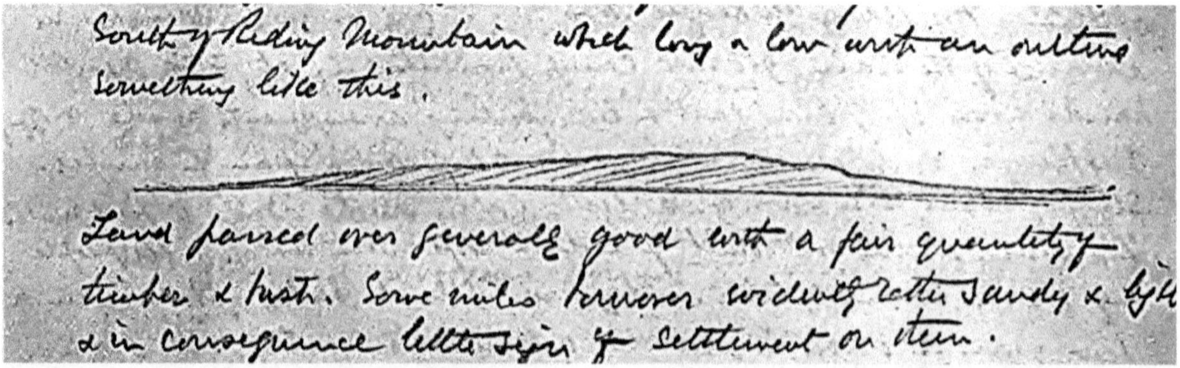

South of Riding Mountain which long a low with an outline something like this.

Land passed over generally good with a fair quantity of timber & brush. Some miles however evidently rather sandy & light & in consequence little sign of settlement on them.

Dawson 1879. McGill 1022-6-135 p258

Dawson 1879. Geological Survey of Canada field notebooks.

Dawson 1881. McGill University 1022-6-81 p37

Dawson 1881. McGill University 1022-6-81 p23

Dawson c. 1881. McGill University 1022-6-81 p55

Dawson 1881. McGill University 1022-6-137 p158

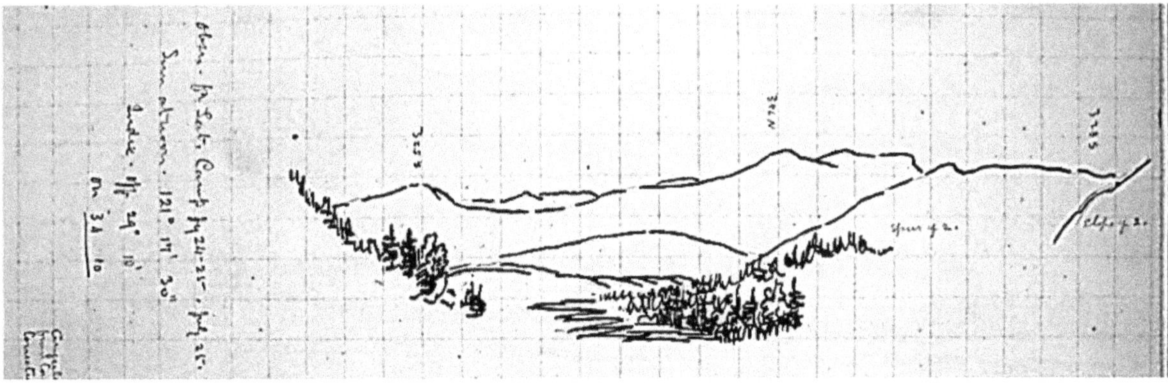

Dawson 1883. Geological Survey of Canada field notebooks.

Dawson 1883. Geological Survey of Canada field notebooks.

Dawson 1884. Geological Survey of Canada field notebooks.

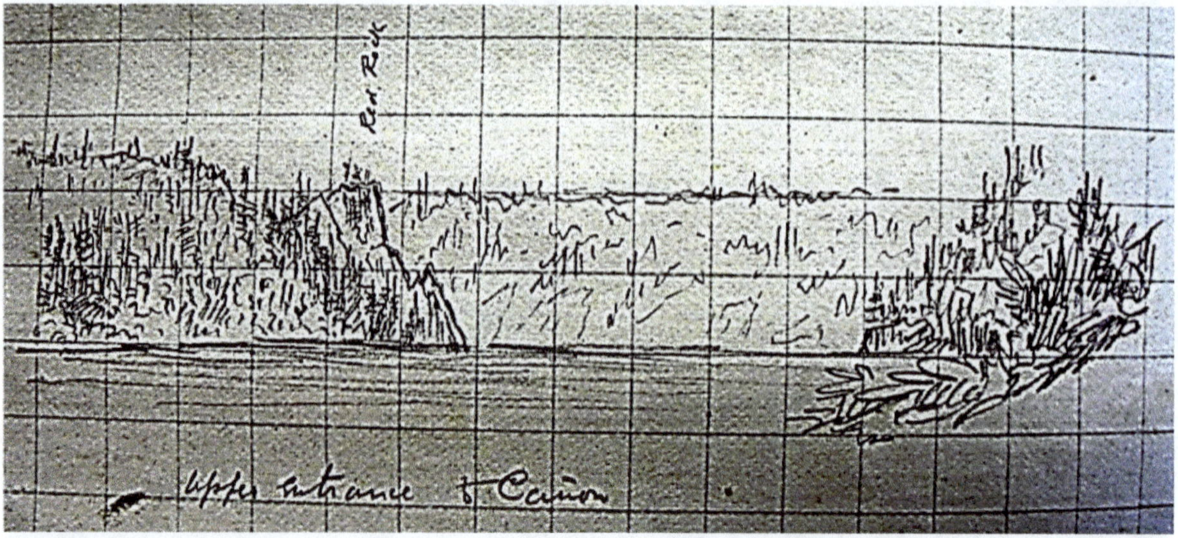

Dawson 1887. Geological Survey of Canada field notebooks.

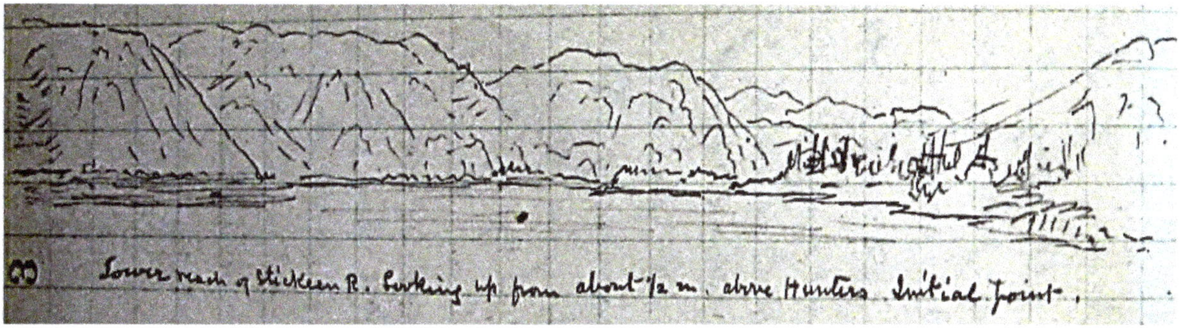

Dawson 1884. Geological Survey of Canada field notebooks.

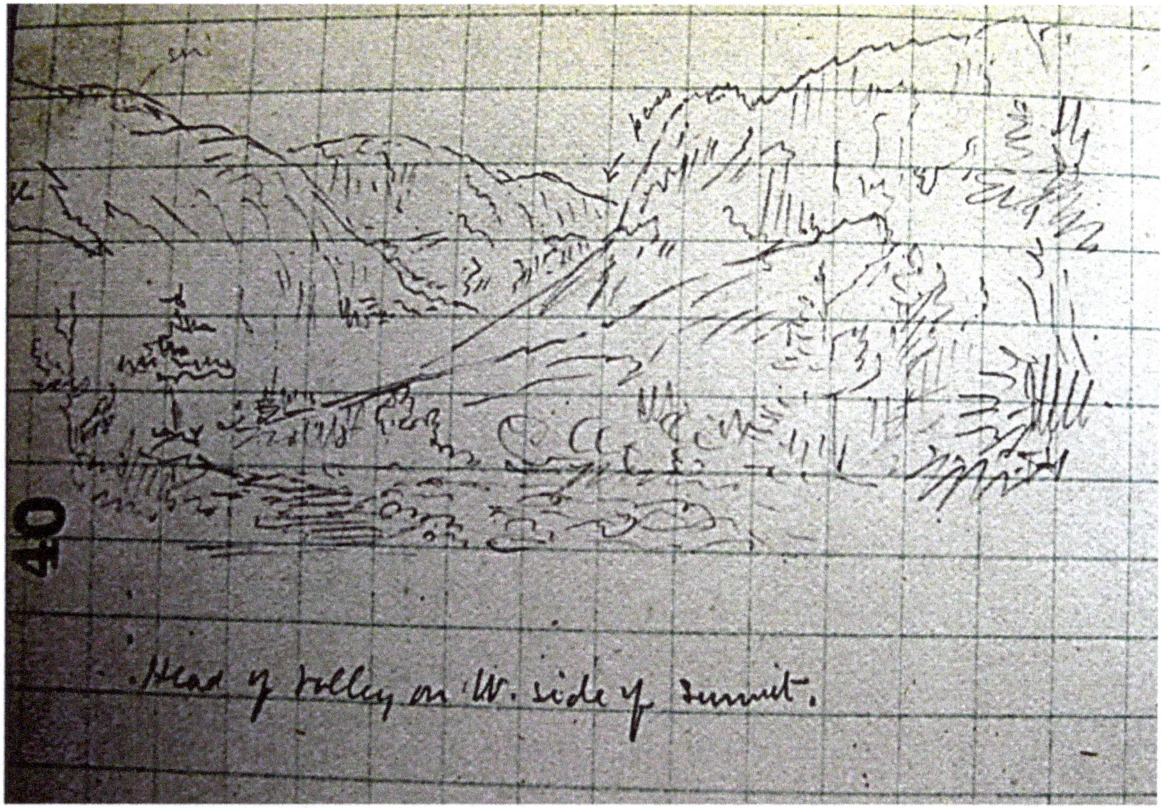

Dawson 1887. Geological Survey of Canada field notebooks.

Dawson 1887. Geological Survey of Canada field notebooks.

Dawson 1888. Geological Survey of Canada field notebooks.

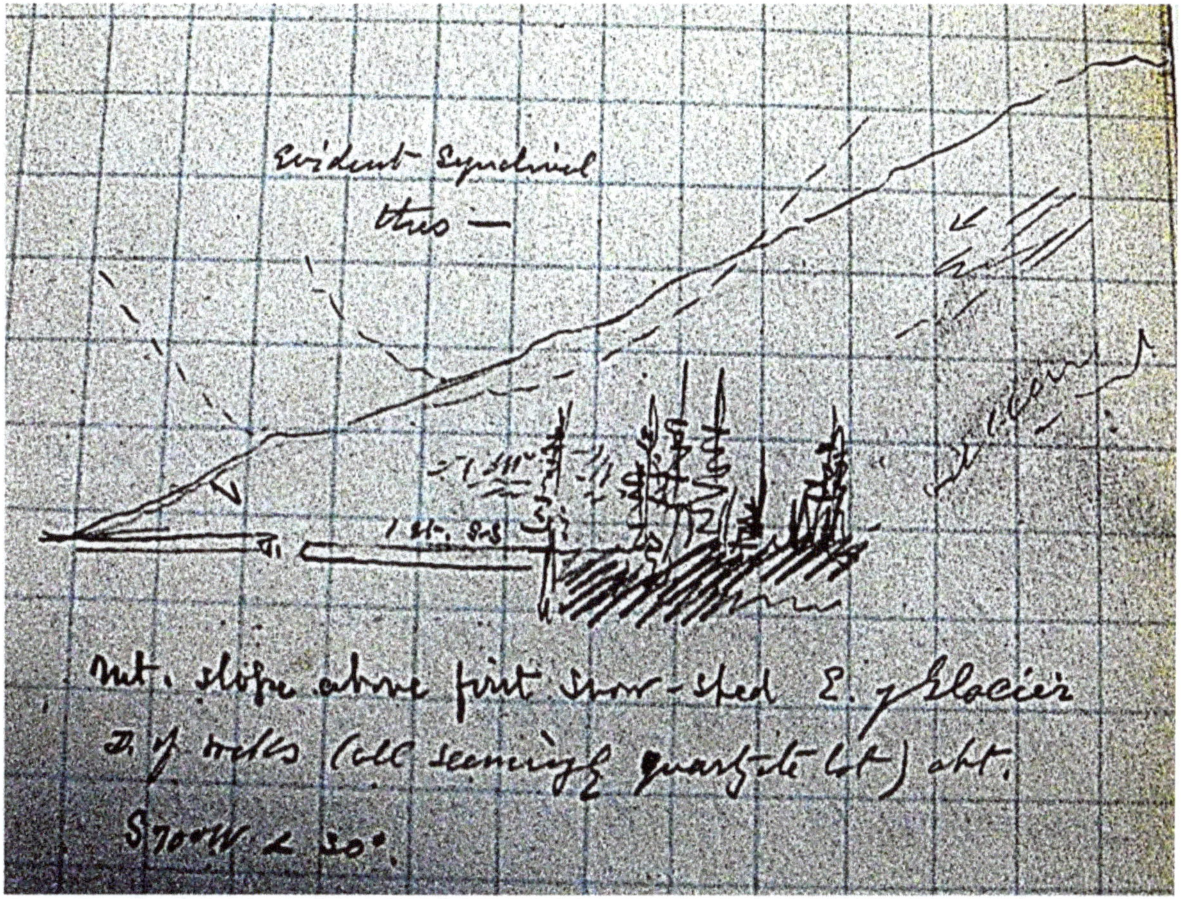

Dawson 1890. Geological Survey of Canada field notebooks.

Dawson 1891. Sketch A.
McGill University 1022-6-64 p48

Dawson 1891. Sketch B.
McGill University 1022-6-64 p49

Dawson 1891. Sketch C.
McGill University 1022-6-64 p50

Dawson 1898.
McGill University 1022-6-156 p192

Elevations

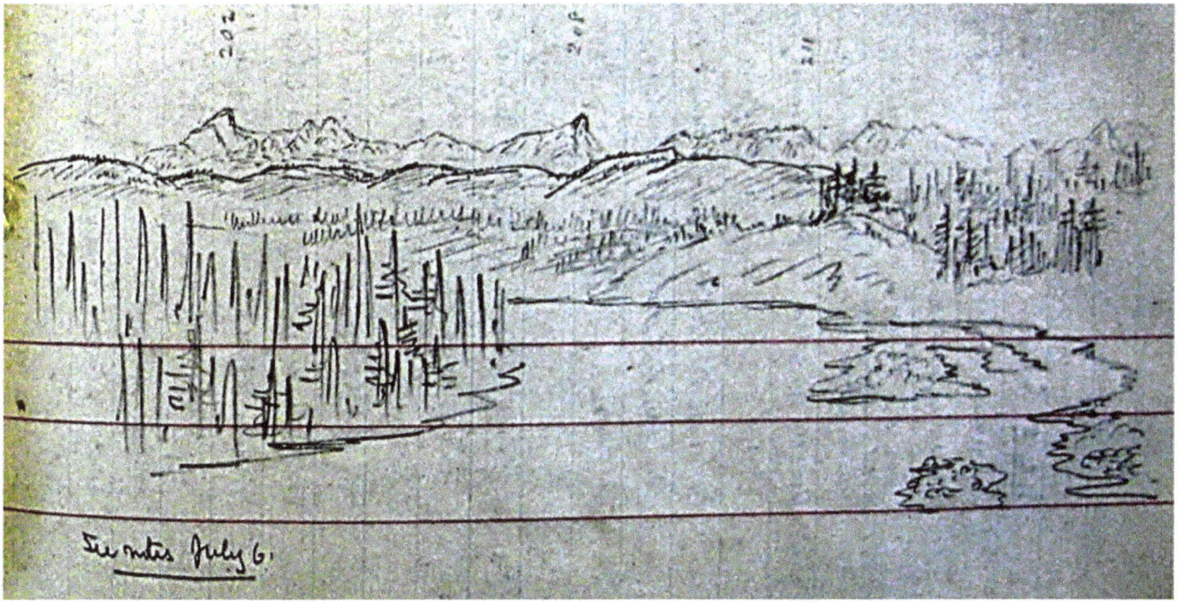

Dawson 1875. Geological Survey of Canada field notebooks.

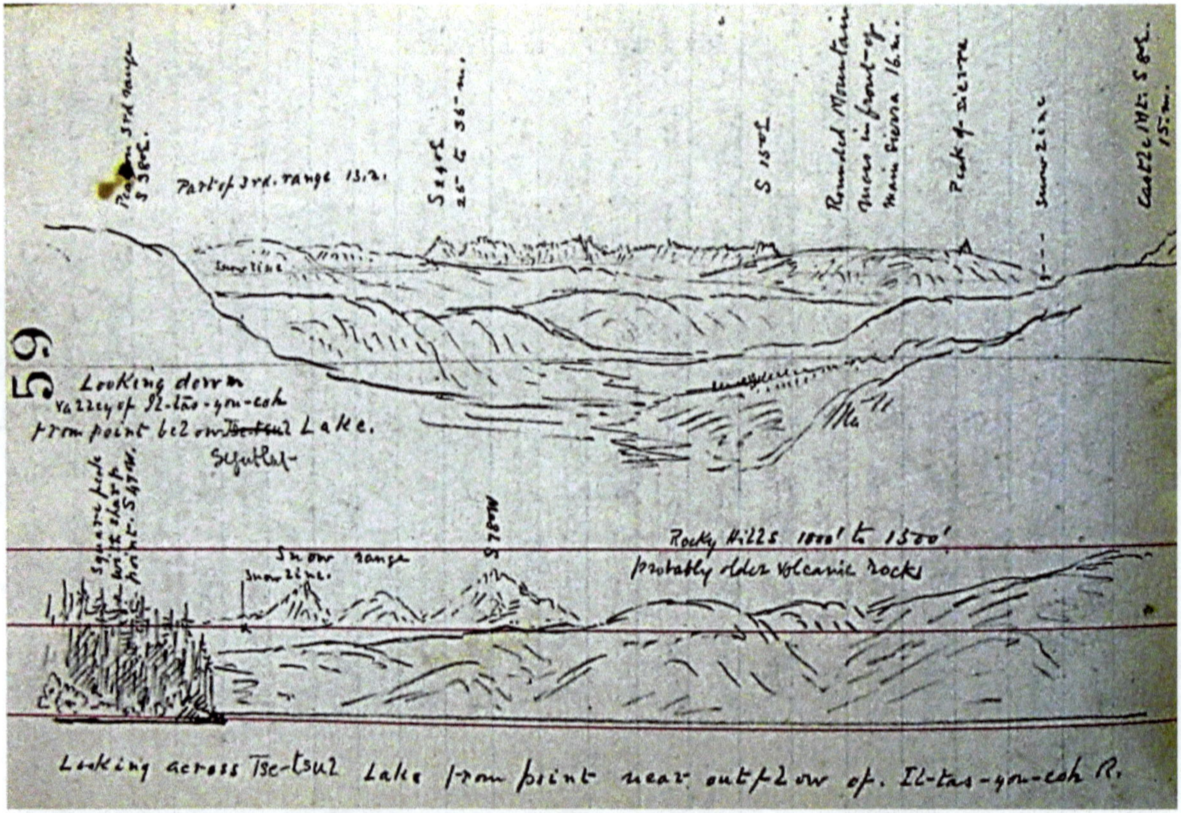

Dawson 1875. Geological Survey of Canada field notebooks.

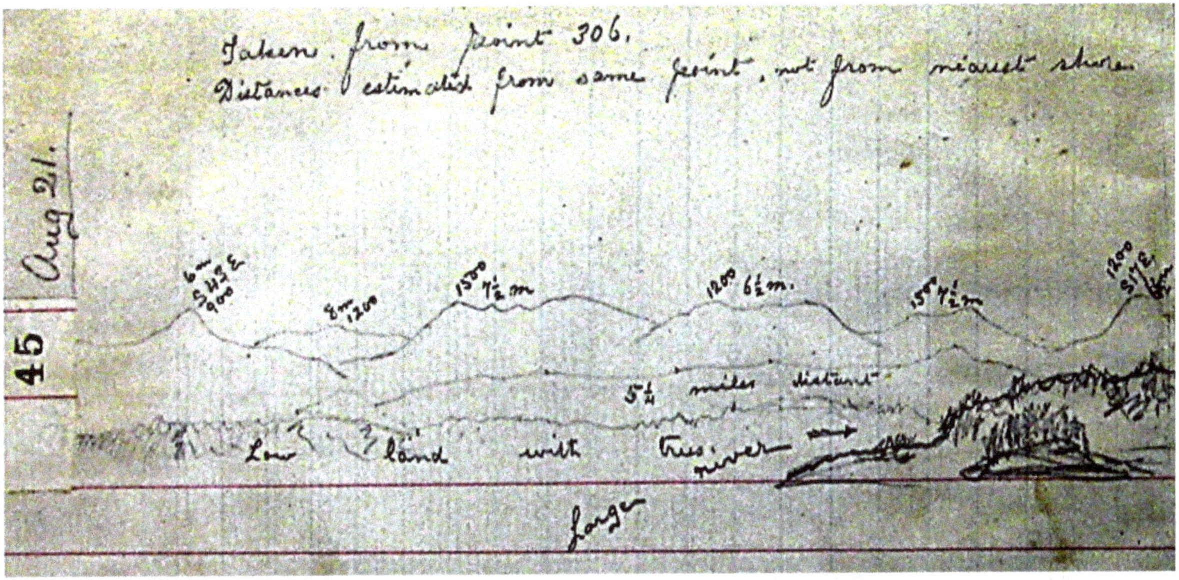

Dawson 1875. Geological Survey of Canada field notebooks. LEFT

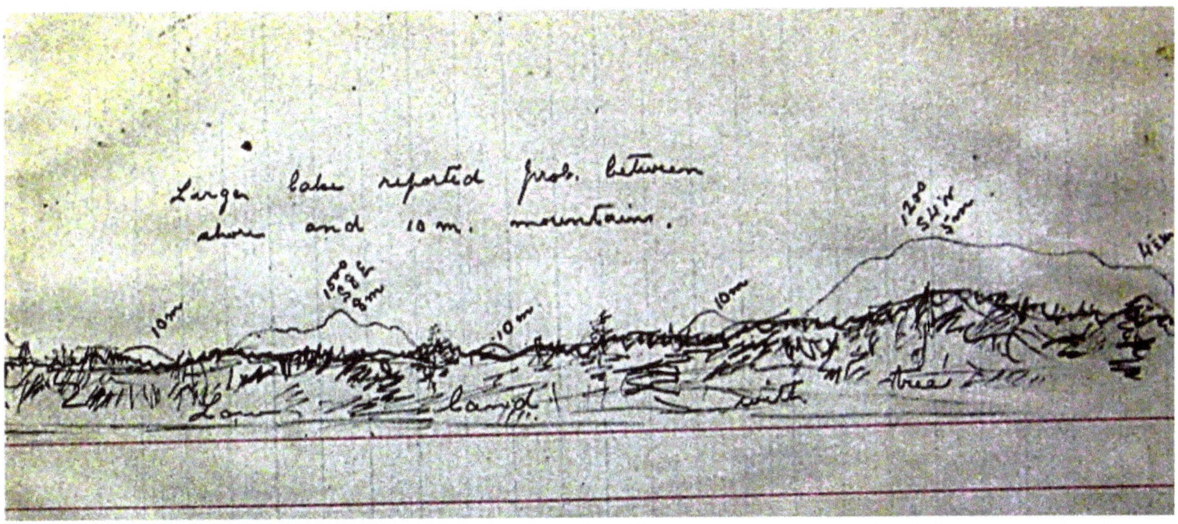

Dawson 1875. Geological Survey of Canada field notebooks. RIGHT

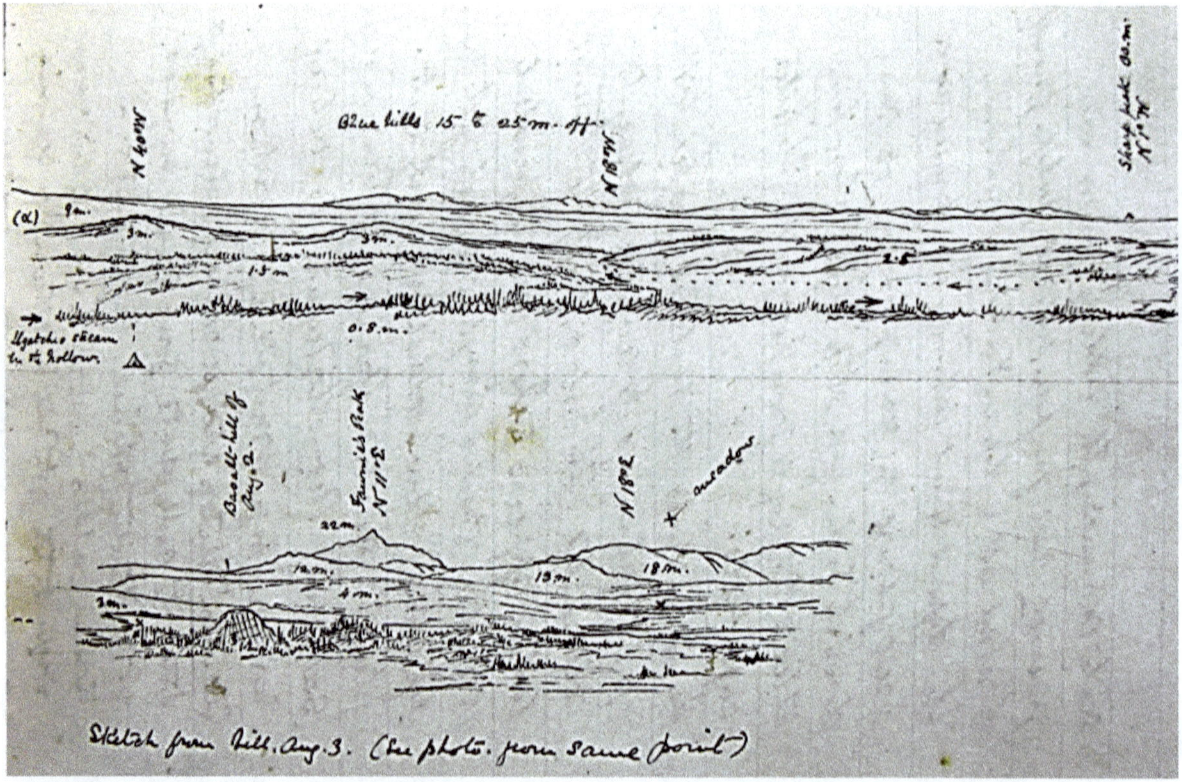

Dawson 1876. Geological Survey of Canada field notebooks.

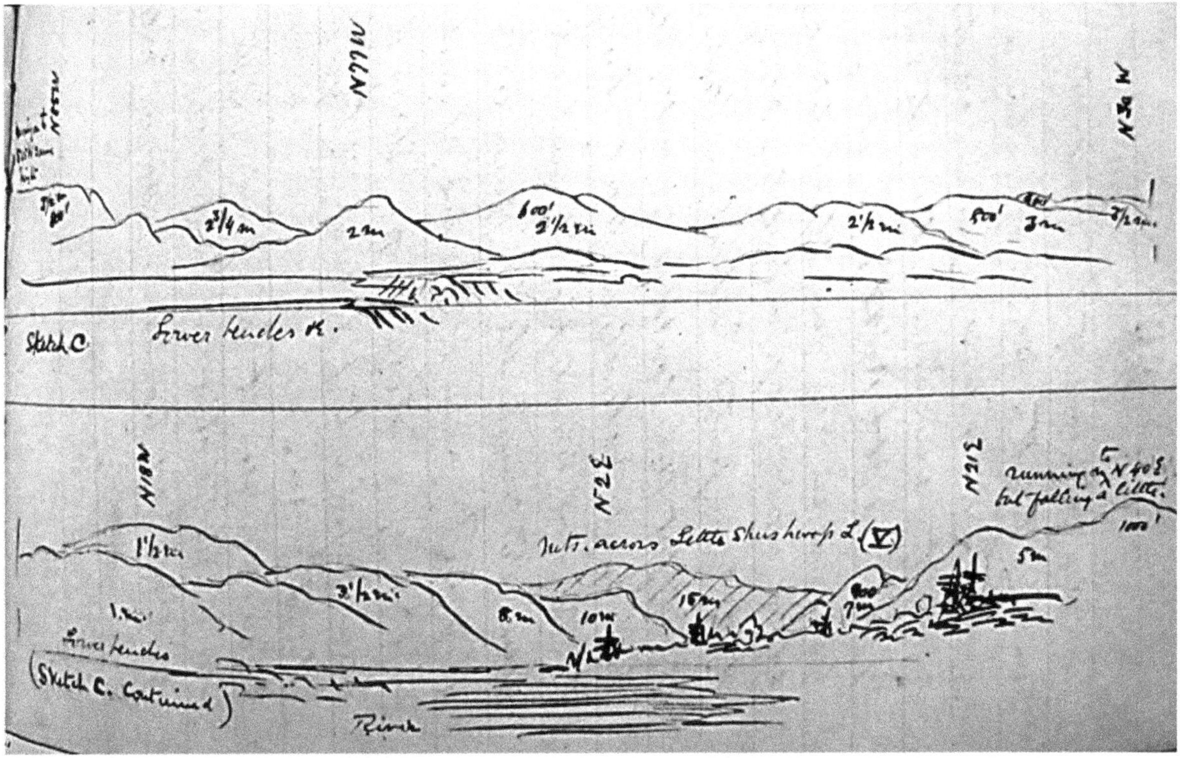

Dawson 1877. Geological Survey of Canada field notebooks.

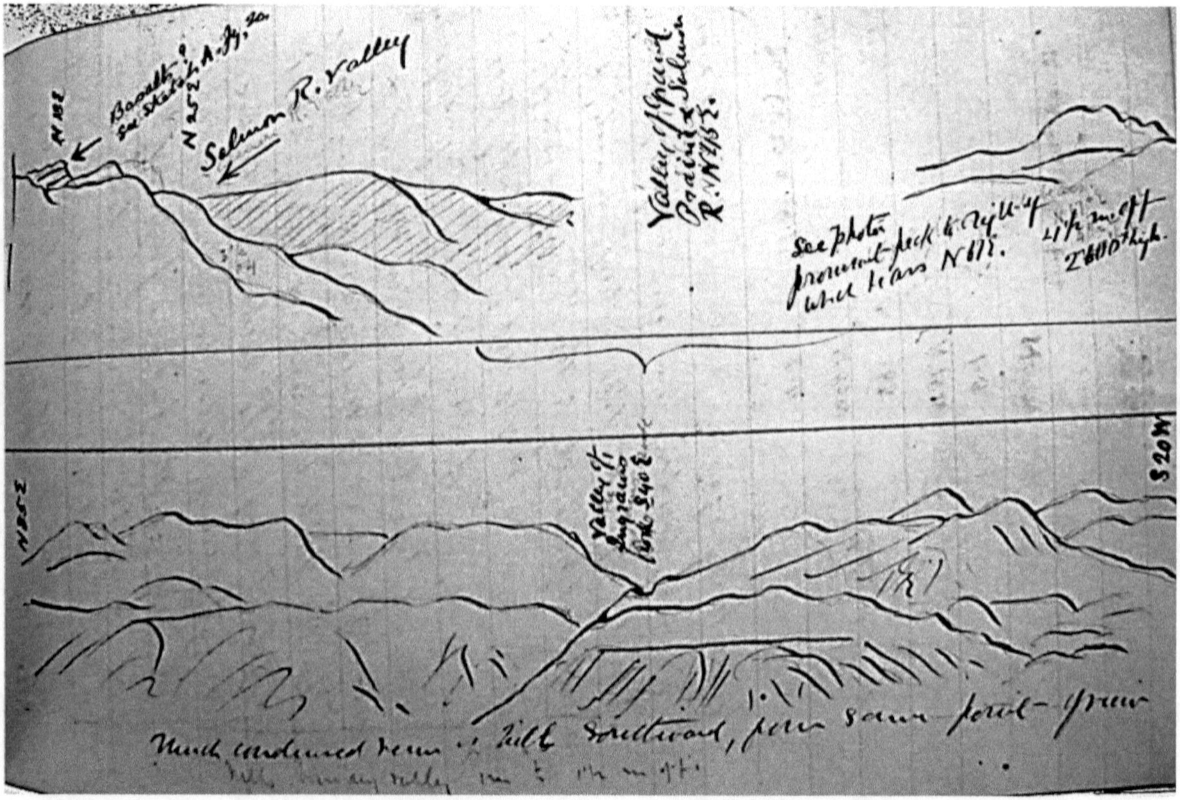

Dawson 1877. Geological Survey of Canada field notebooks.

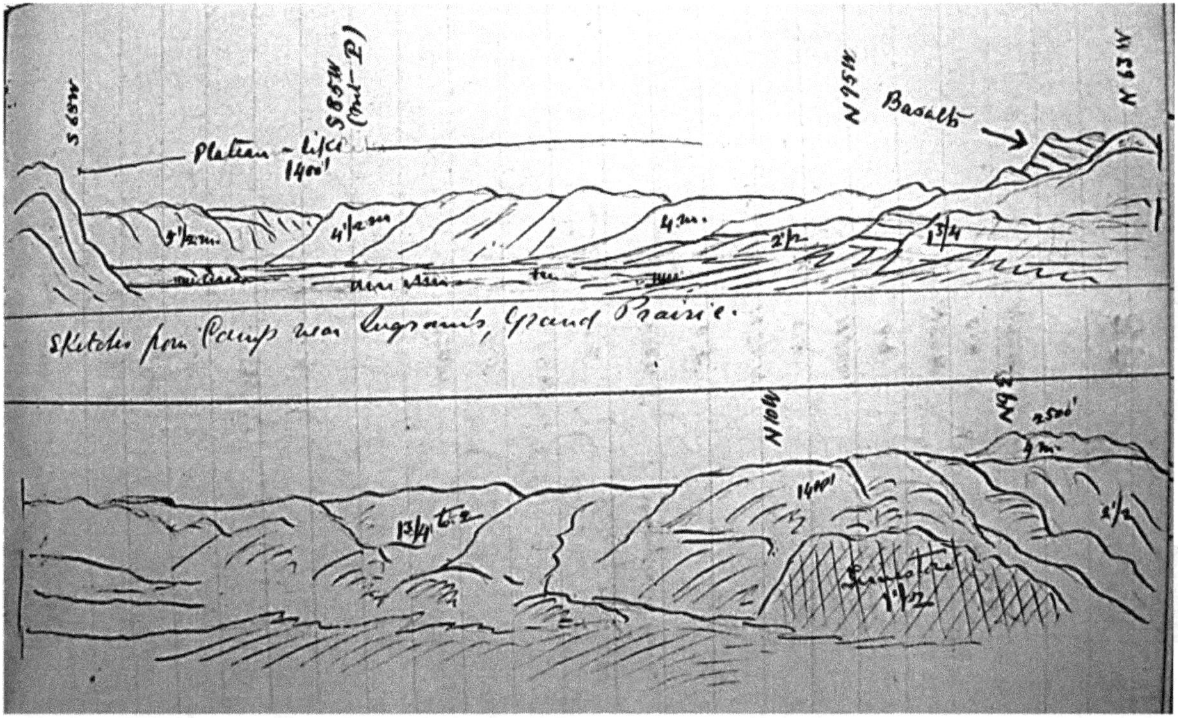

Dawson 1877. Geological Survey of Canada field notebooks.

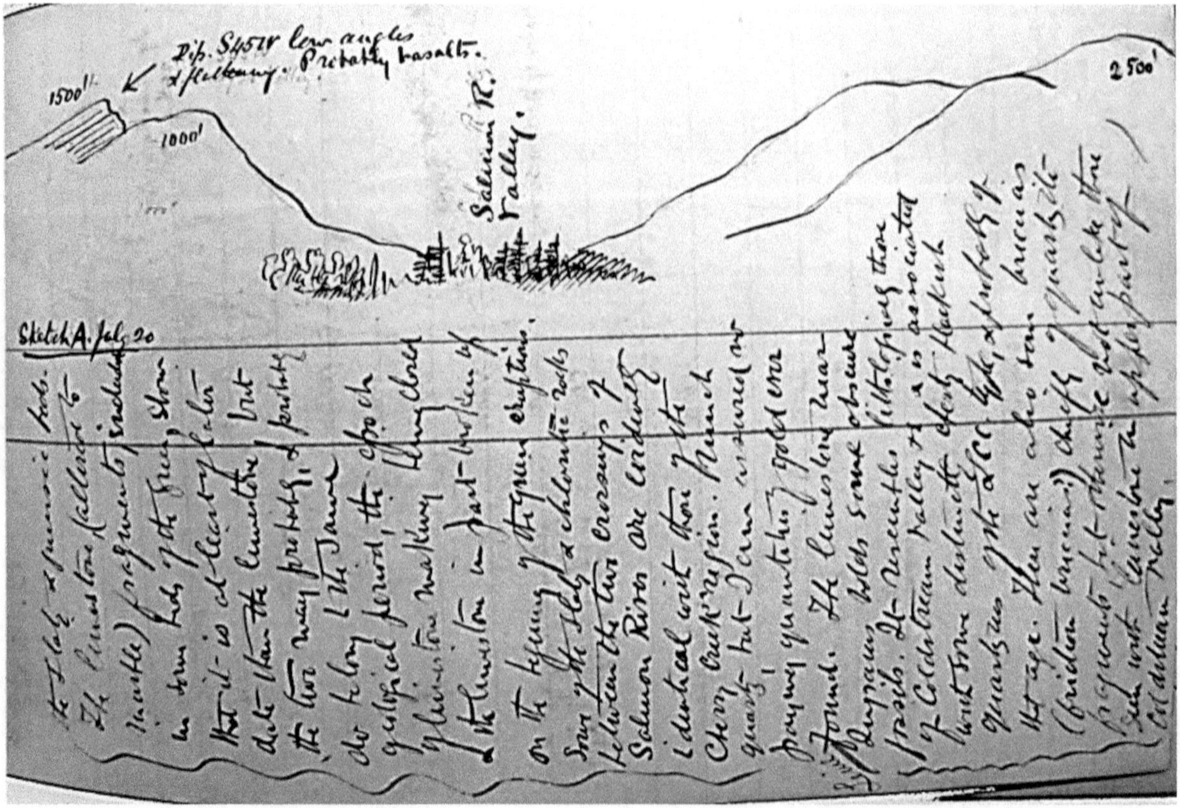

Dawson 1877. Geological Survey of Canada field notebooks.

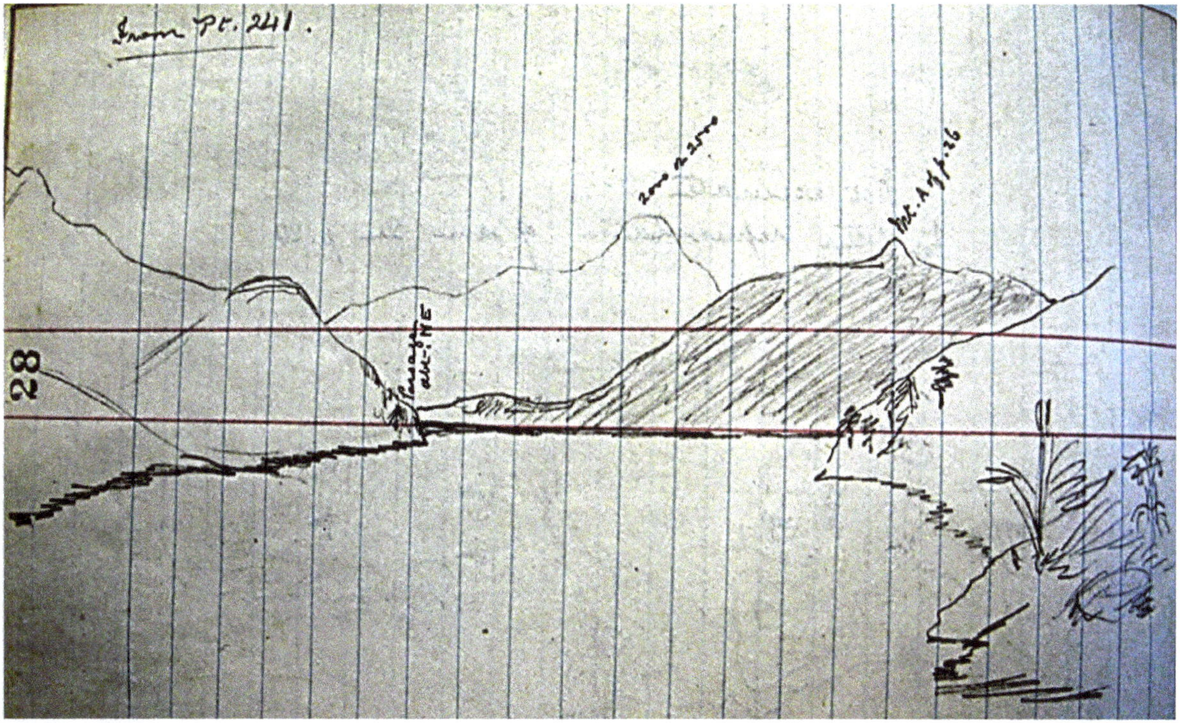

Dawson 1877. Geological Survey of Canada field notebooks.

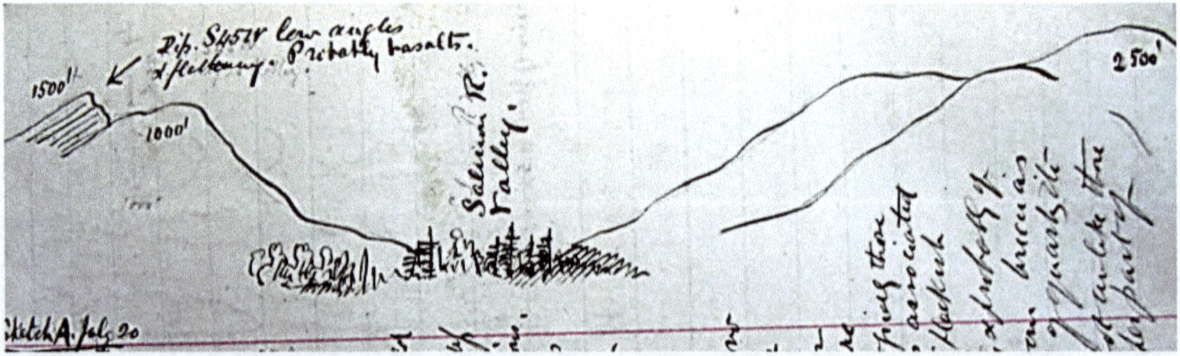

Dawson 1877. Geological Survey of Canada field notebooks.

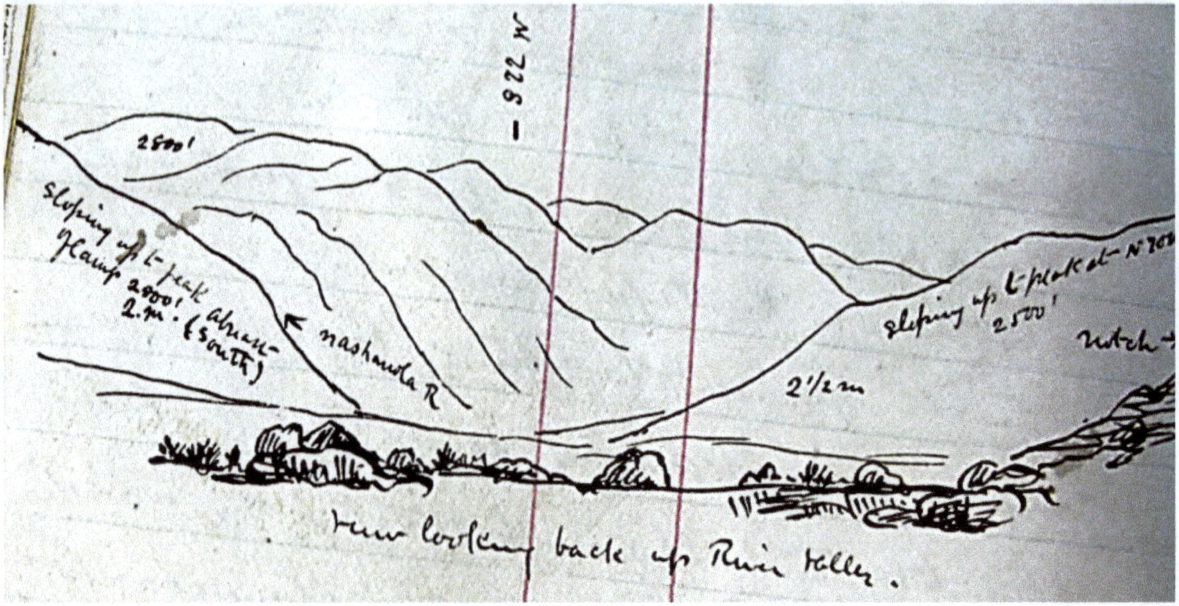

Dawson 1877. Geological Survey of Canada field notebooks.

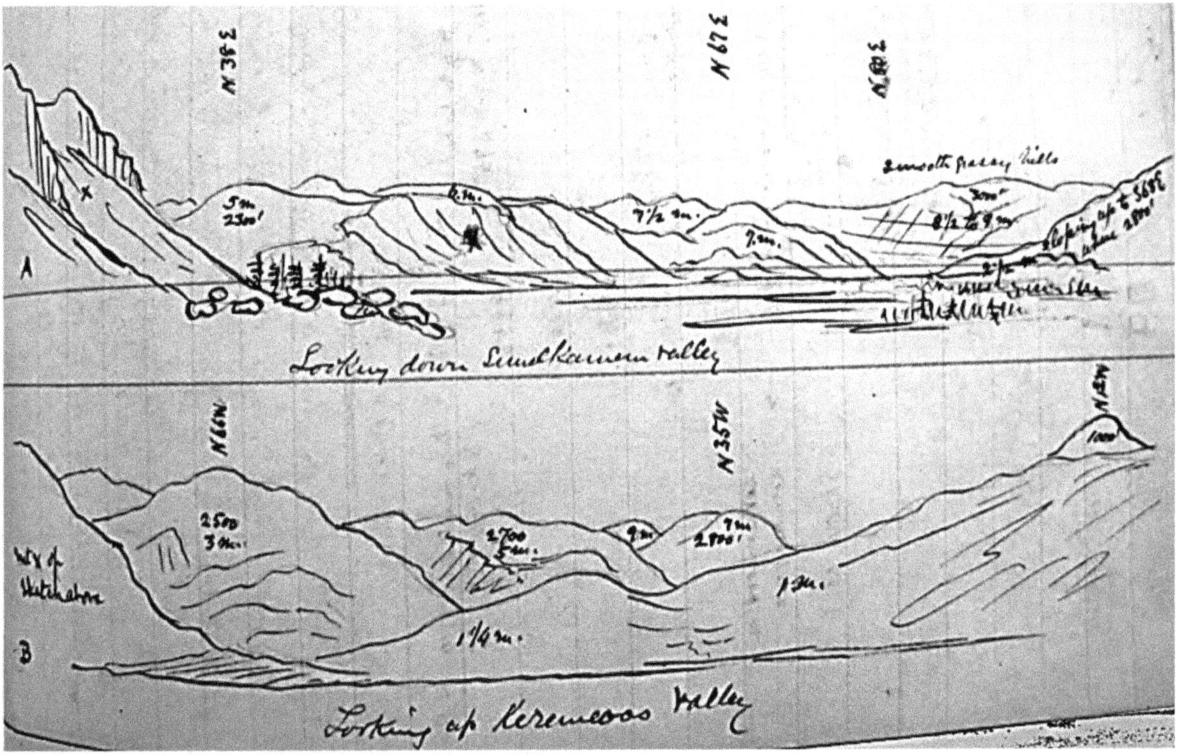

Dawson 1877. Geological Survey of Canada field notebooks.

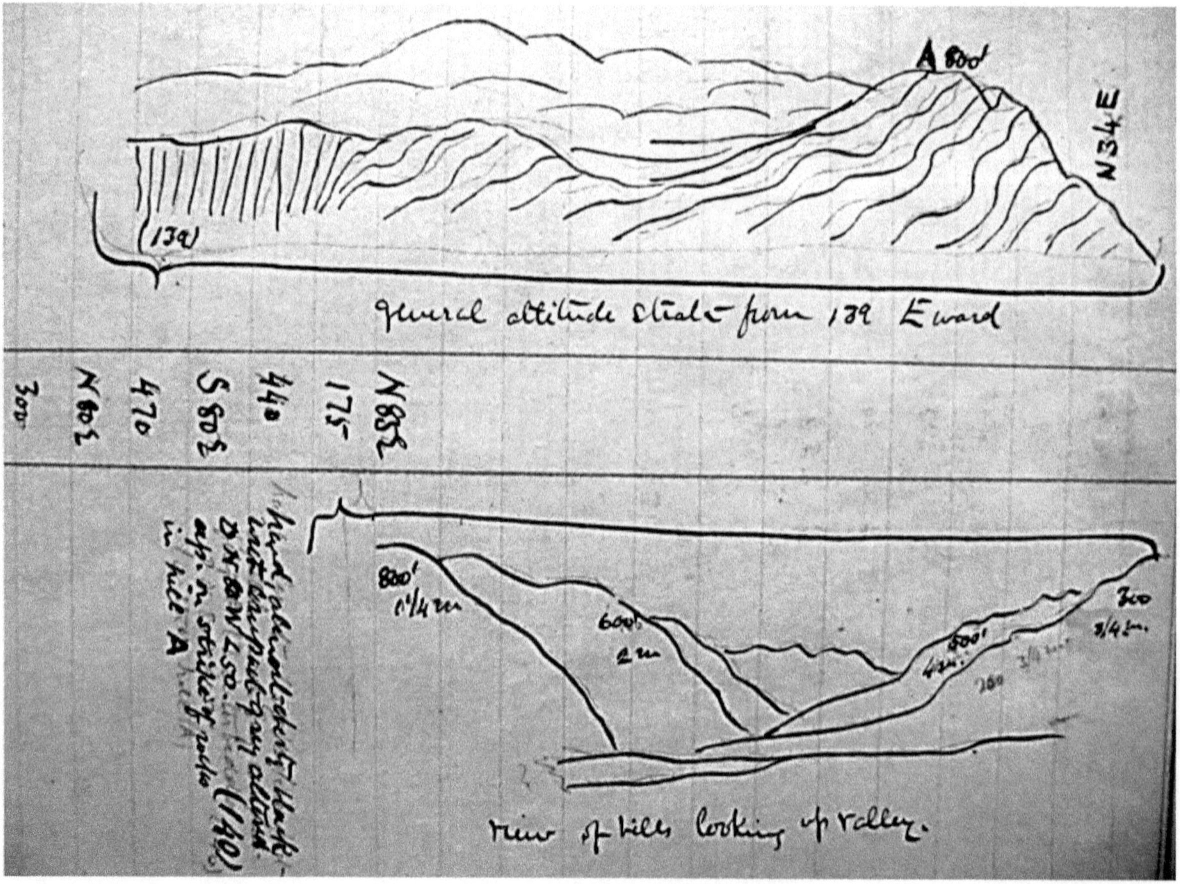

Dawson 1877. Geological Survey of Canada field notebooks.

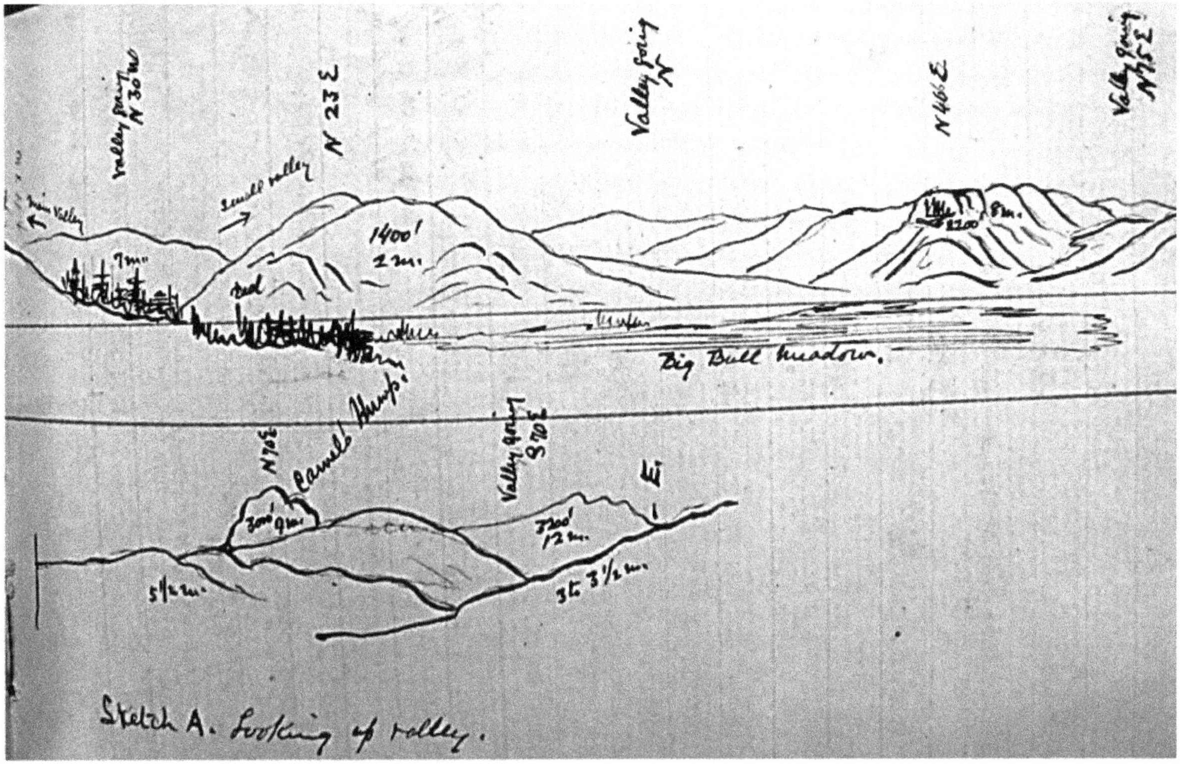

Dawson 1877. Geological Survey of Canada field notebooks.

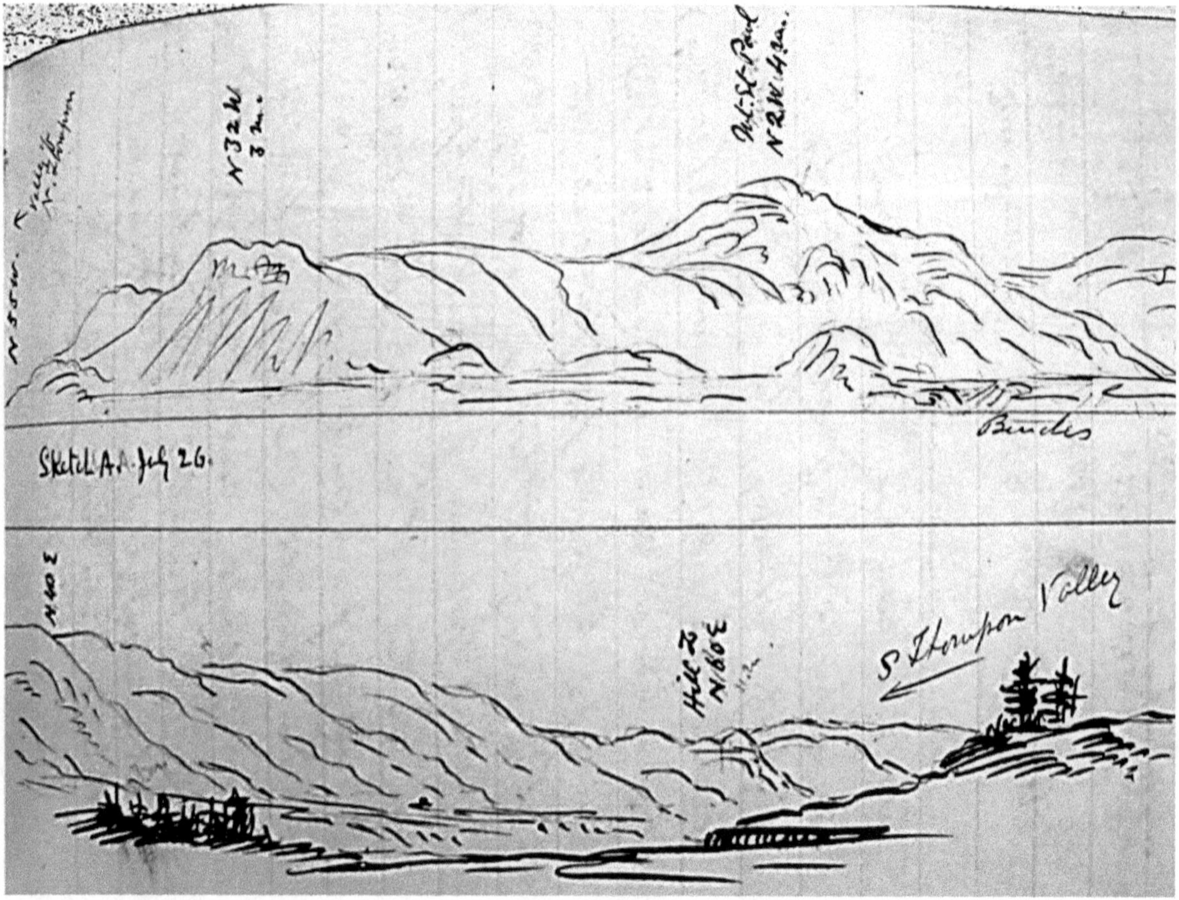

Dawson 1877. Geological Survey of Canada field notebooks.

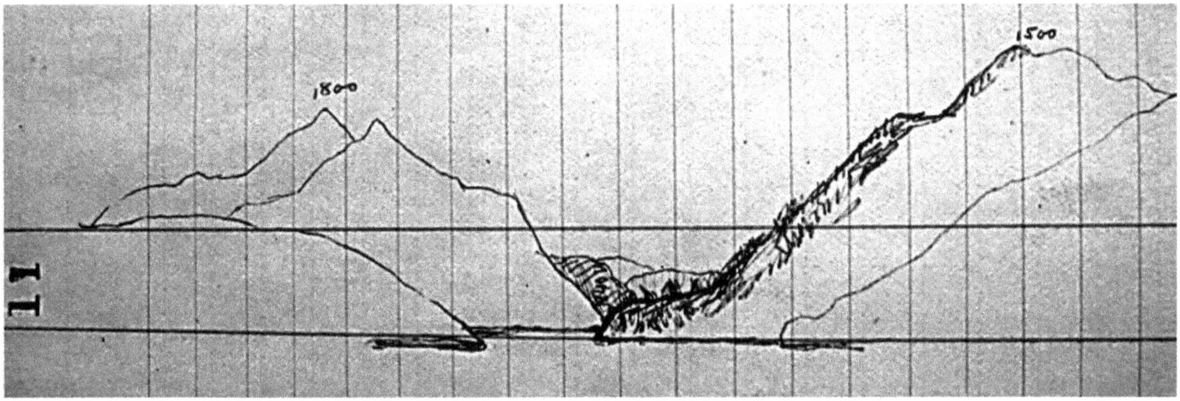

Dawson 1877. Geological Survey of Canada field notebooks.

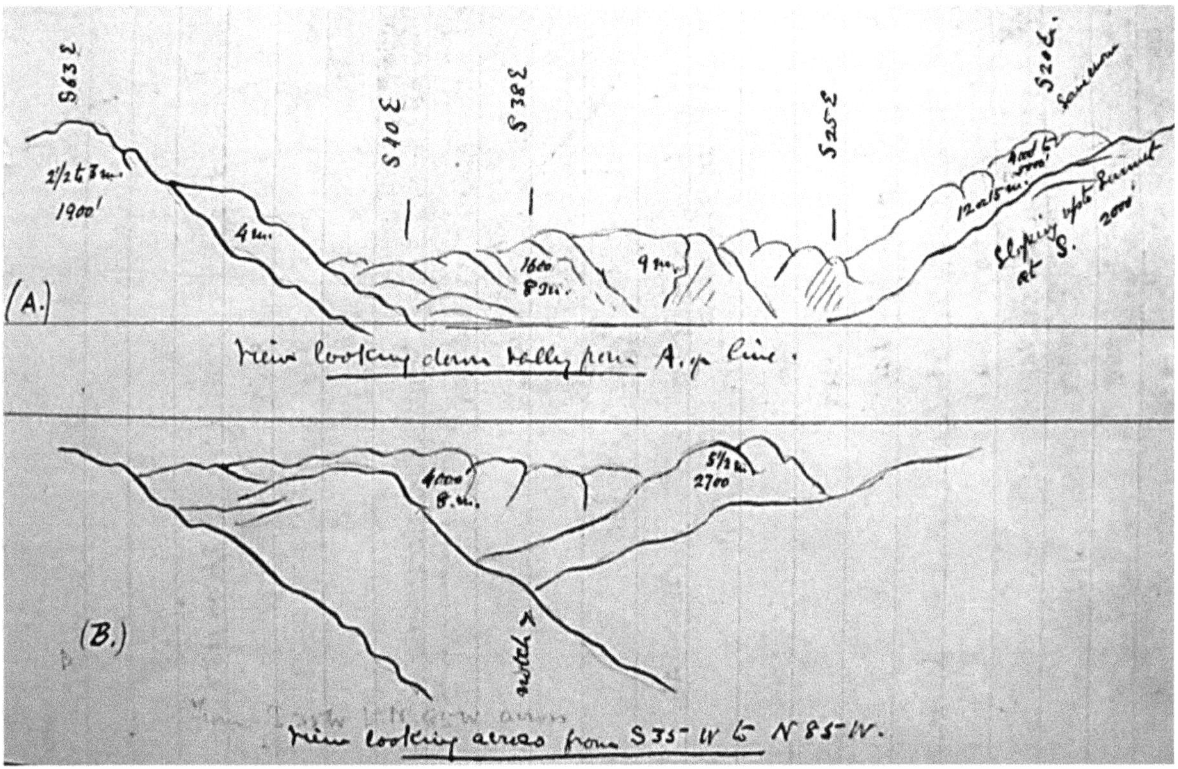

Dawson 1878. Geological Survey of Canada field notebooks.

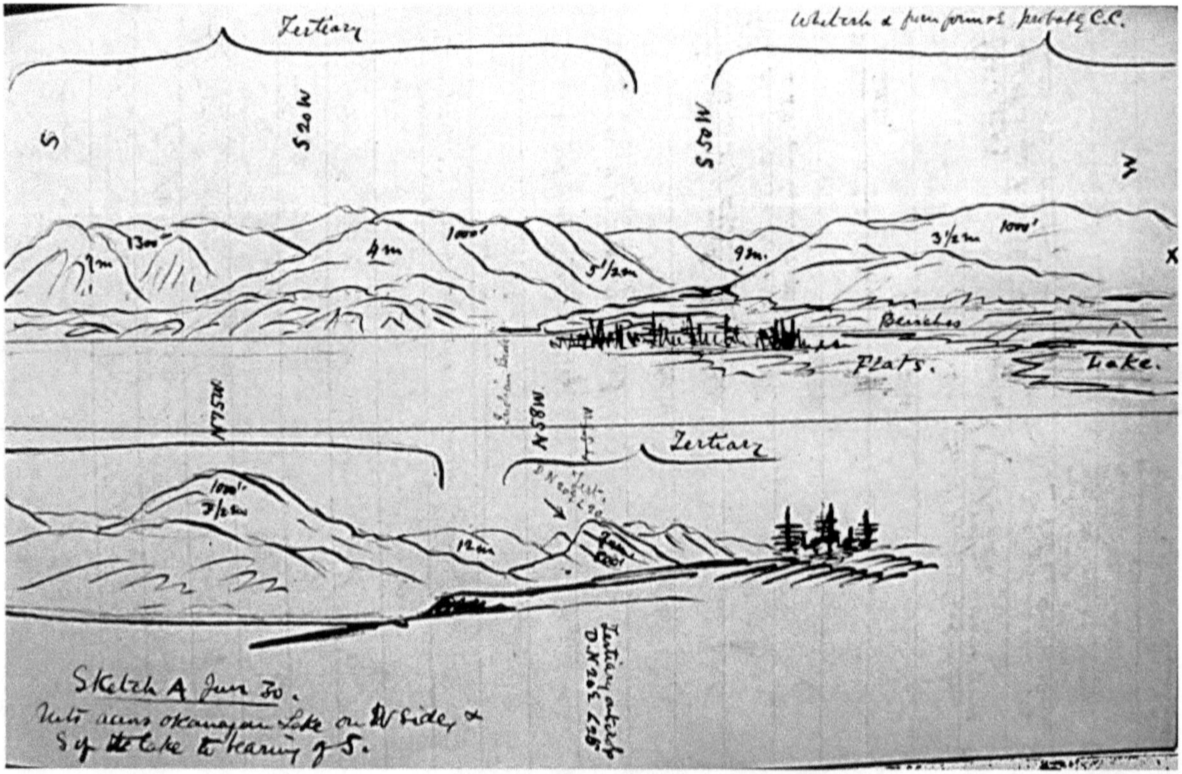

Dawson 1878. Geological Survey of Canada field notebooks.

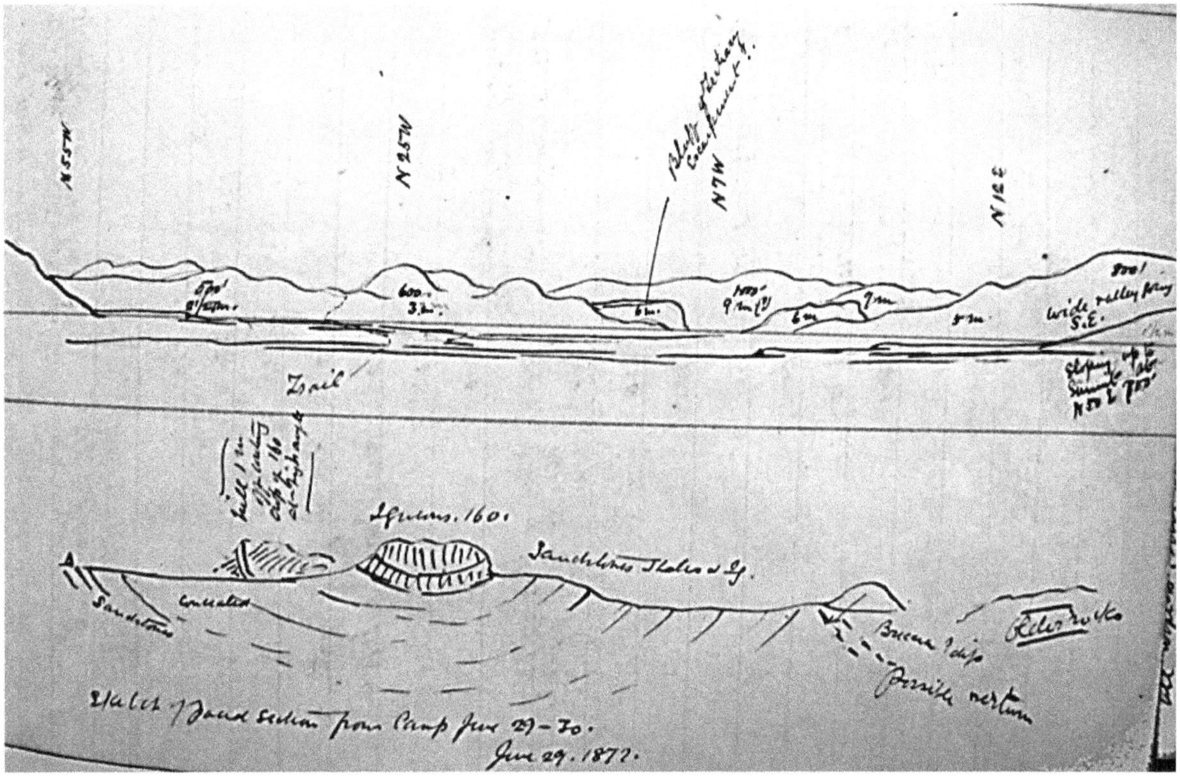

Dawson 1878. Geological Survey of Canada field notebooks.

Dawson 1878. Geological Survey of Canada field notebooks.

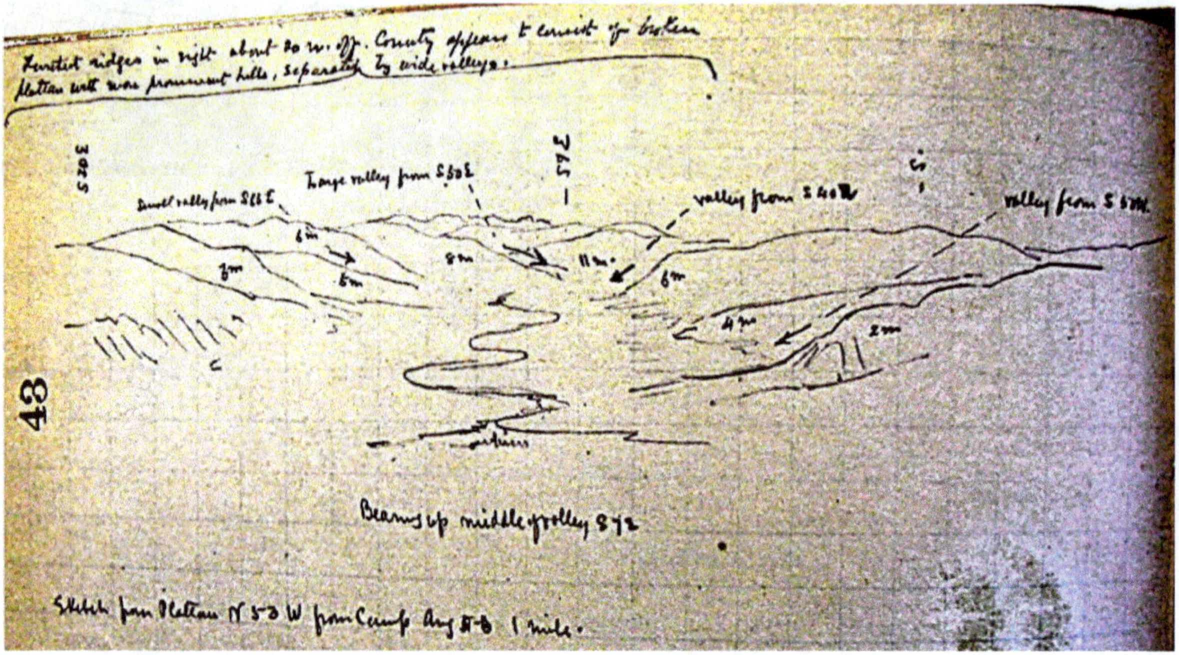

Dawson 1878. Geological Survey of Canada field notebooks.

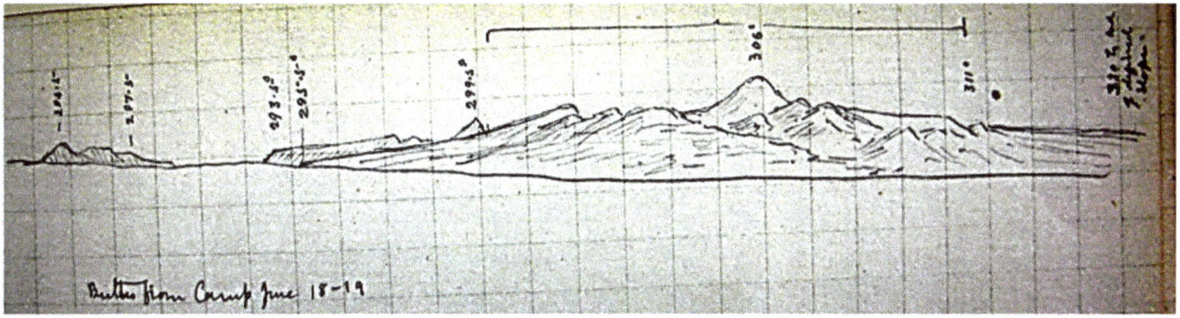

Dawson 1881 no 5. Geological Survey of Canada field notebooks.

Dawson 1879. Geological Survey of Canada field notebooks.

Dawson 1881. Geological Survey of Canada field notebooks.

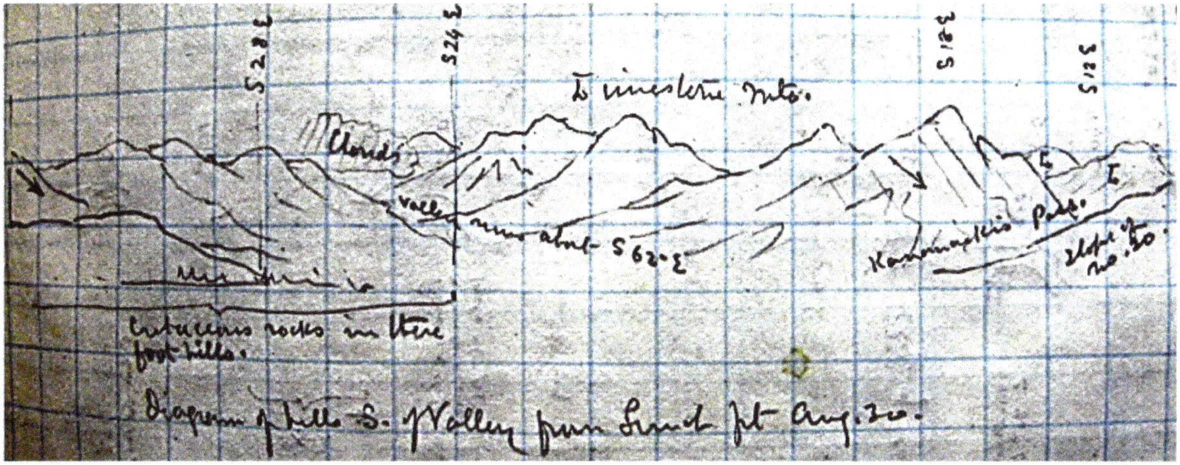

Dawson 1881 no 3.. Geological Survey of Canada field notebooks.

Dawson 1881 no 5. Geological Survey of Canada field notebooks.

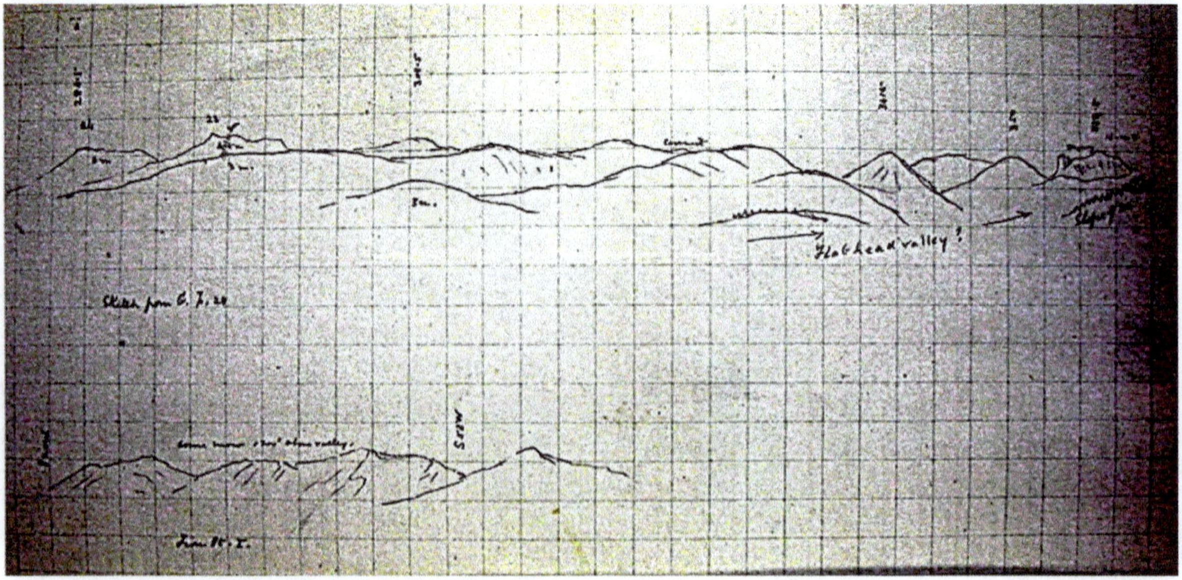

Dawson 1883 no 1. Geological Survey of Canada field notebooks.

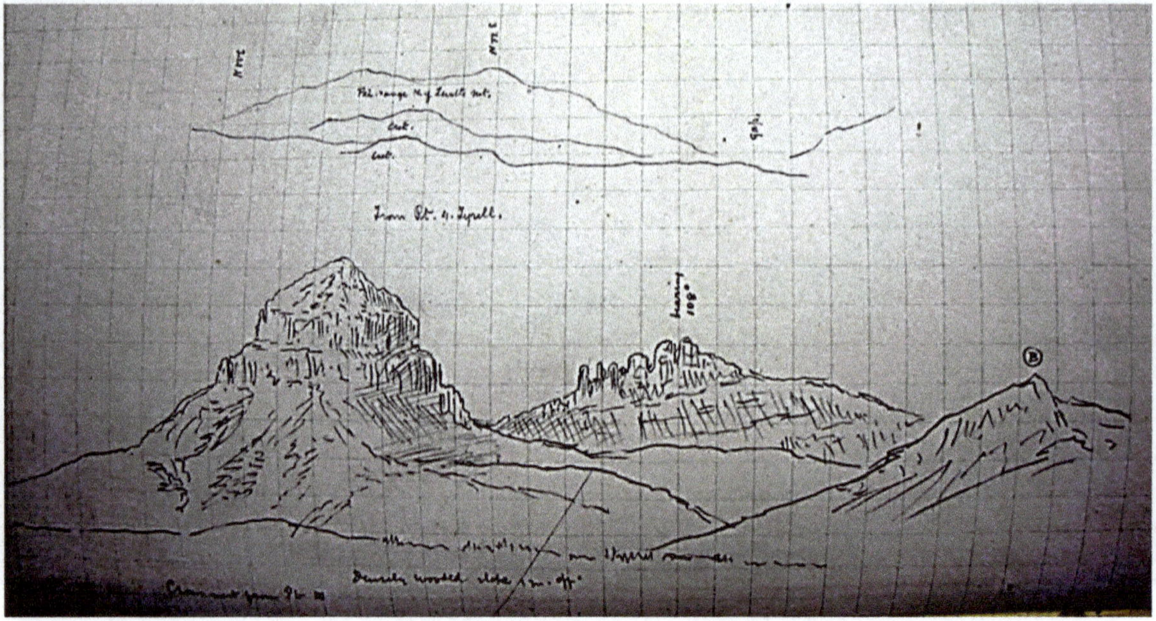

Dawson 1883 no 1. Geological Survey of Canada field notebooks.

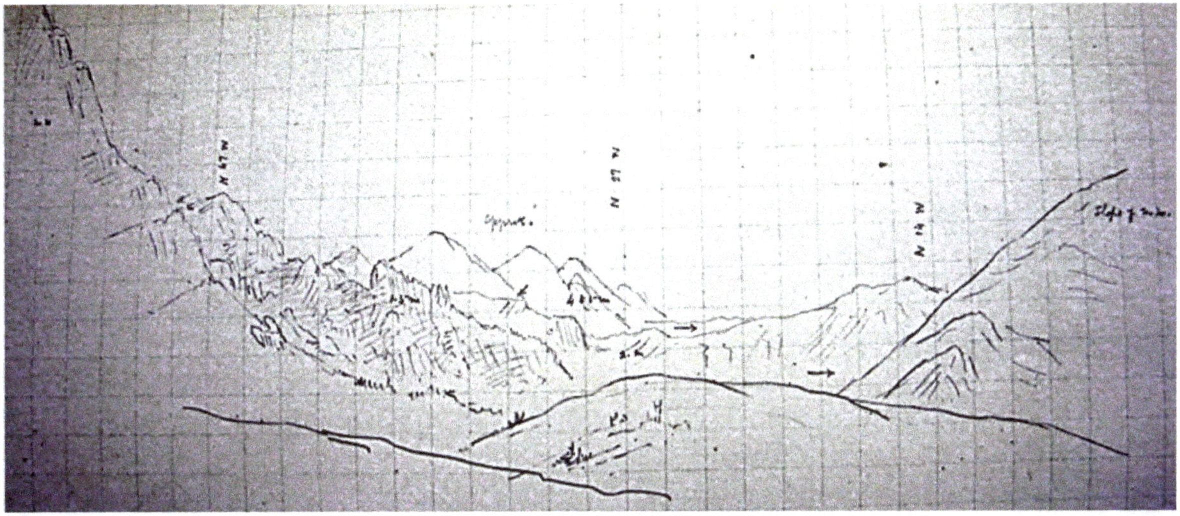

Dawson 1883 no 1. Geological Survey of Canada field notebooks.

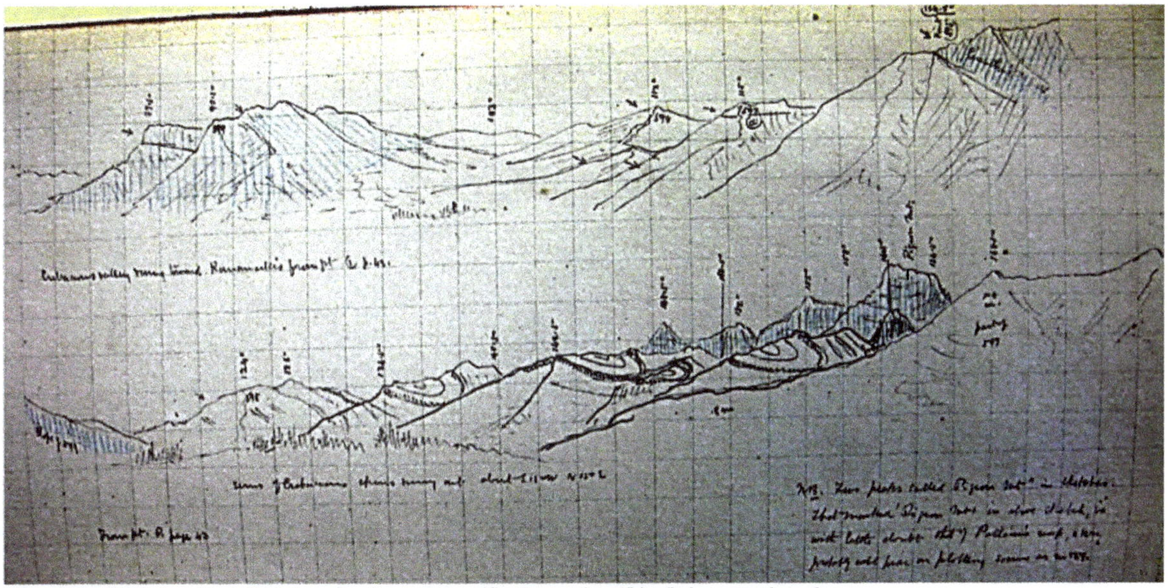

Dawson 1884. Geological Survey of Canada field notebooks.

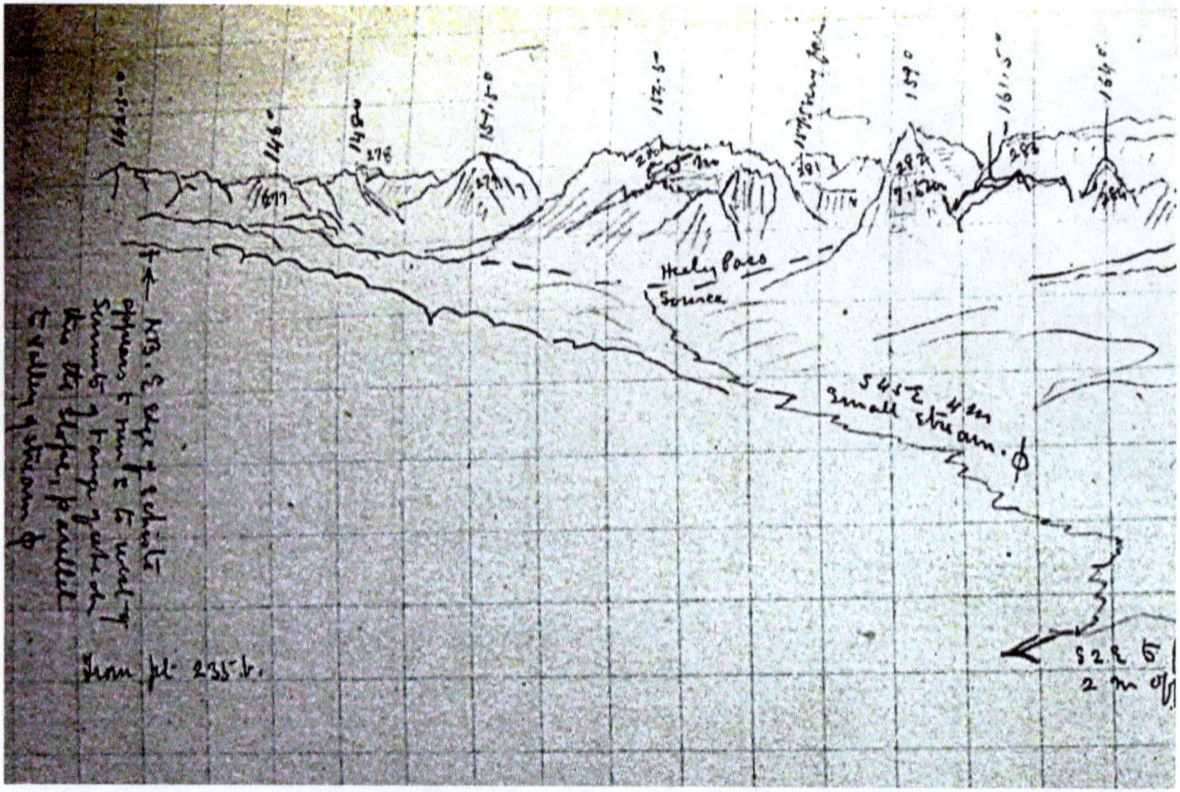

Dawson 1884. Geological Survey of Canada field notebooks. LEFT

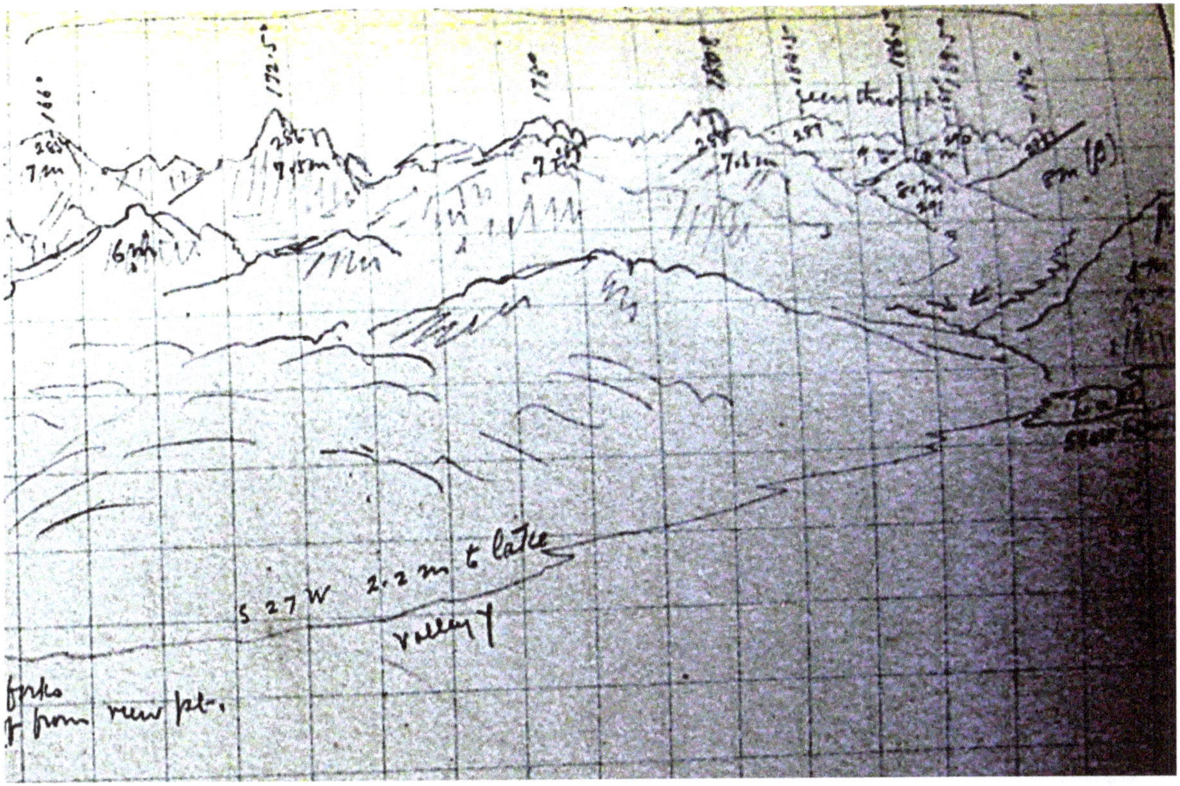

Dawson 1884. Geological Survey of Canada field notebooks. RIGHT

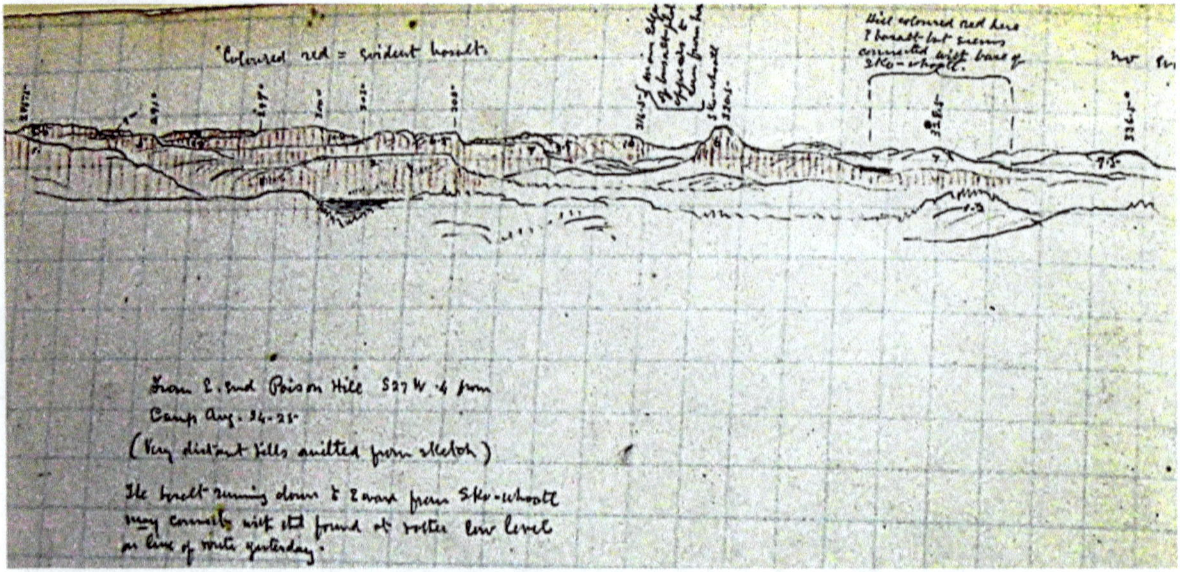

Dawson 1889 no 6. Geological Survey of Canada field notebooks. LEFT

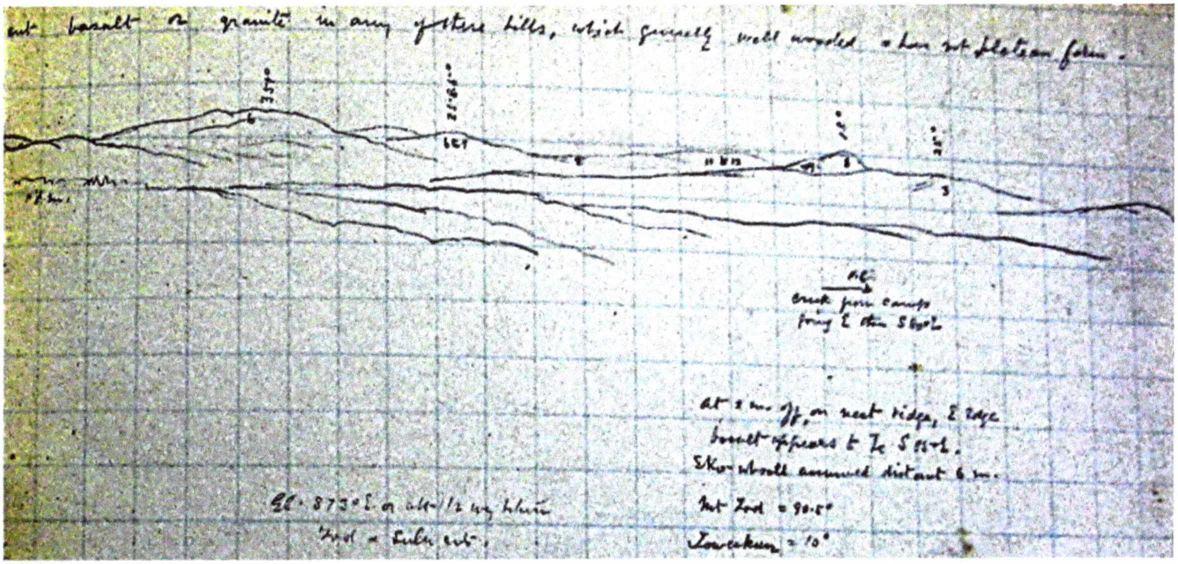

Dawson 1889 no 6. Geological Survey of Canada field notebooks. RIGHT

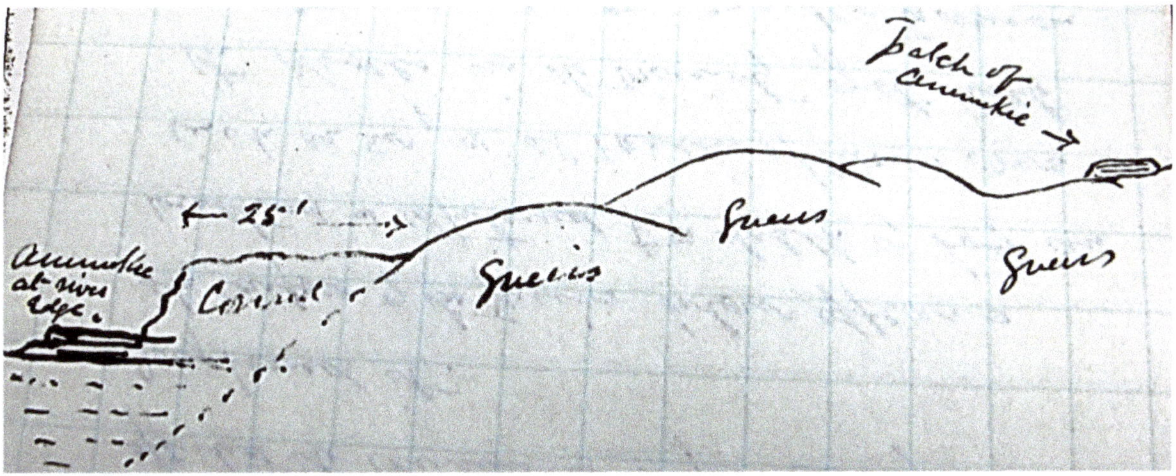

Dawson 1894 no 6. Geological Survey of Canada field notebooks.

Rivers

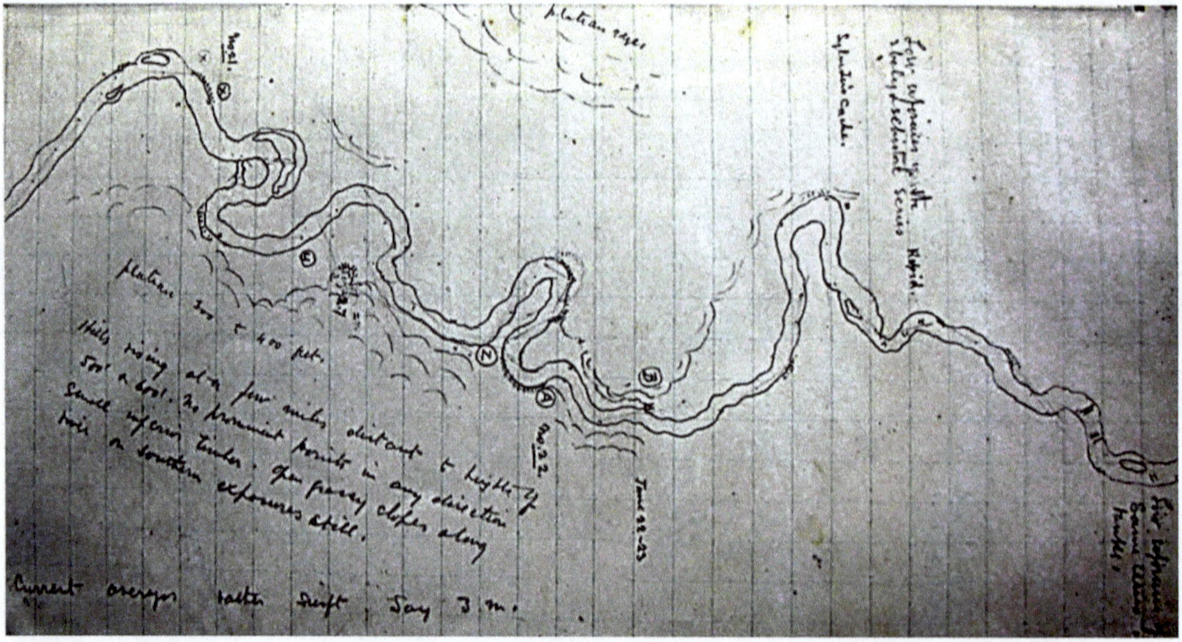

Dawson 1887. Geological Survey of Canada field notebooks.

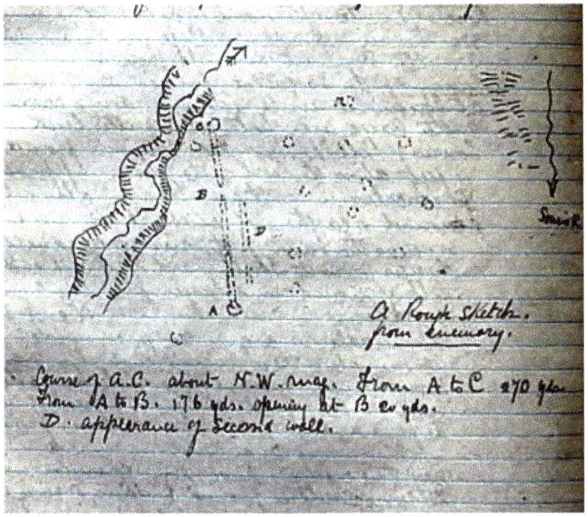

Dawson 1873. McGill niversity
1022 6 177 70

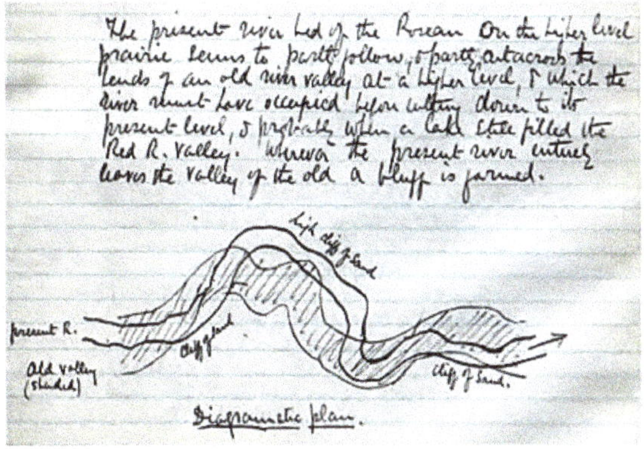

Dawson 1873. McGill niversity
1022 6 178 218 b

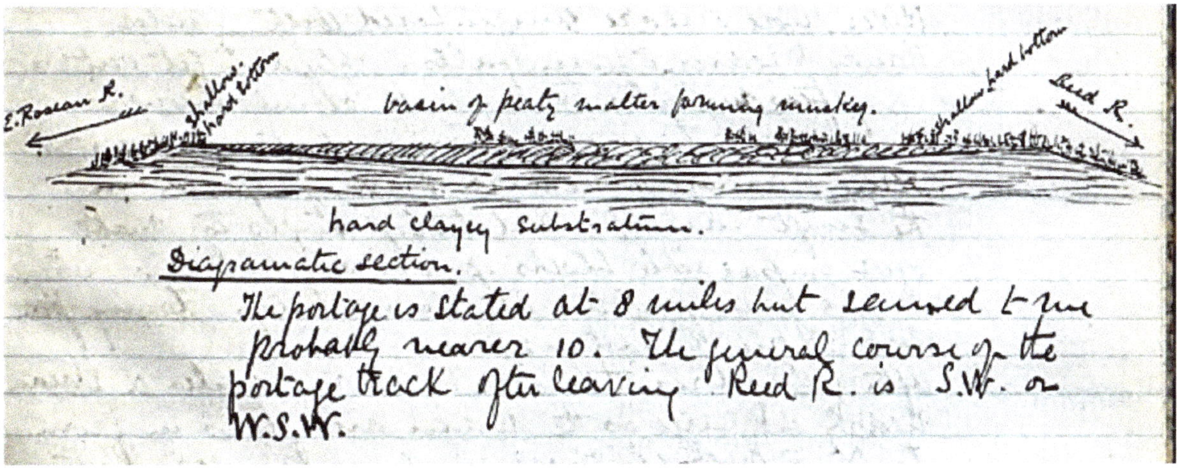

Dawson 1873. McGill niversity 1022 6 178 79

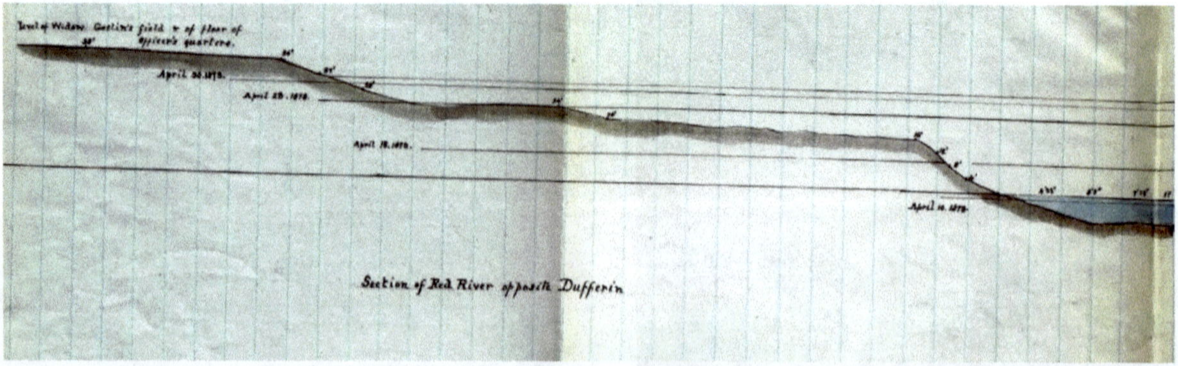

Dawson 1873. Red River.. McGill niversity 1022 6 177 .327 LEFT

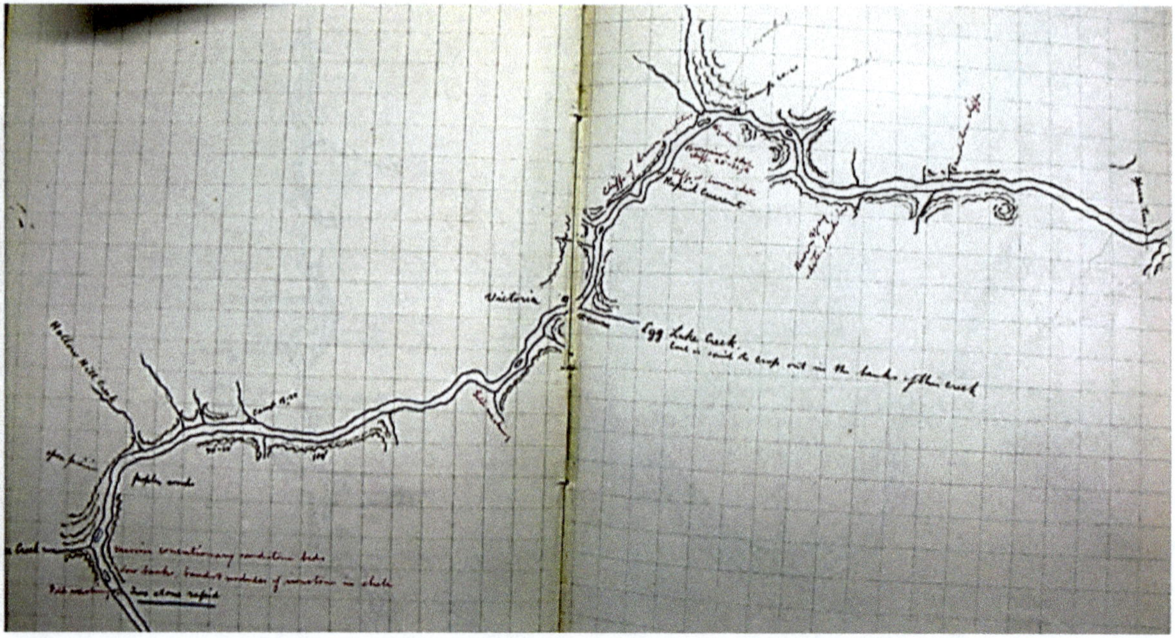

Dawson 1881 no 5. Geological Survey of Canada field notebooks.

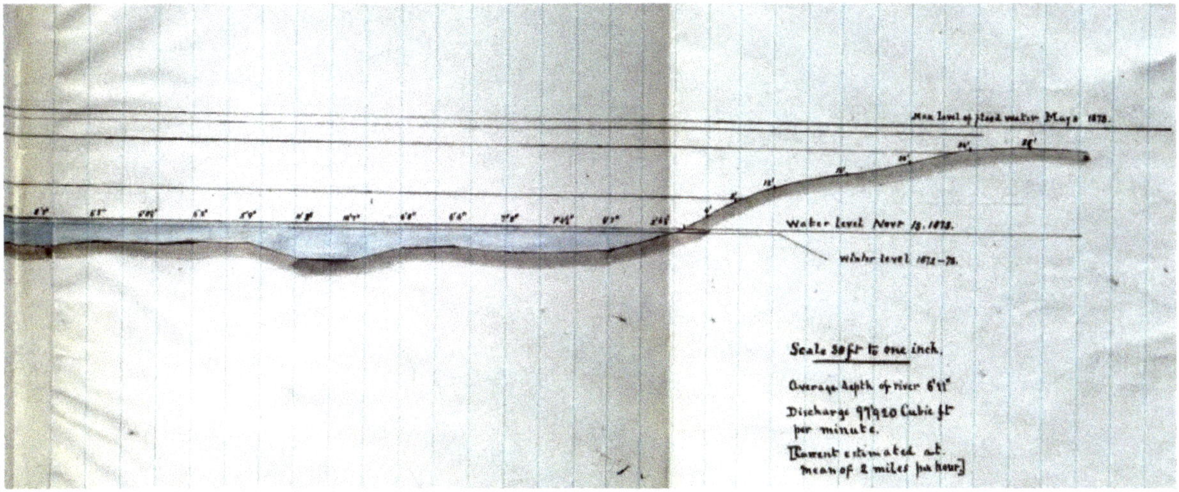

Dawson 1873. Red River.. McGill niversity 1022 6 177 .328 RIGHT

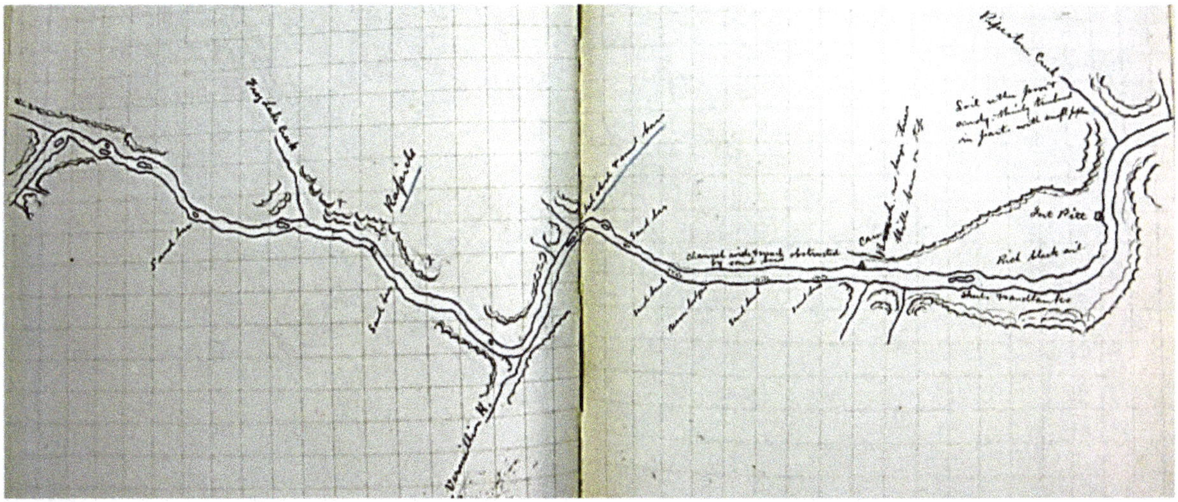

Dawson 1881 no 5. Geological Survey of Canada field notebooks.

Dawson 1878. Geological Survey of Canada field notebooks.

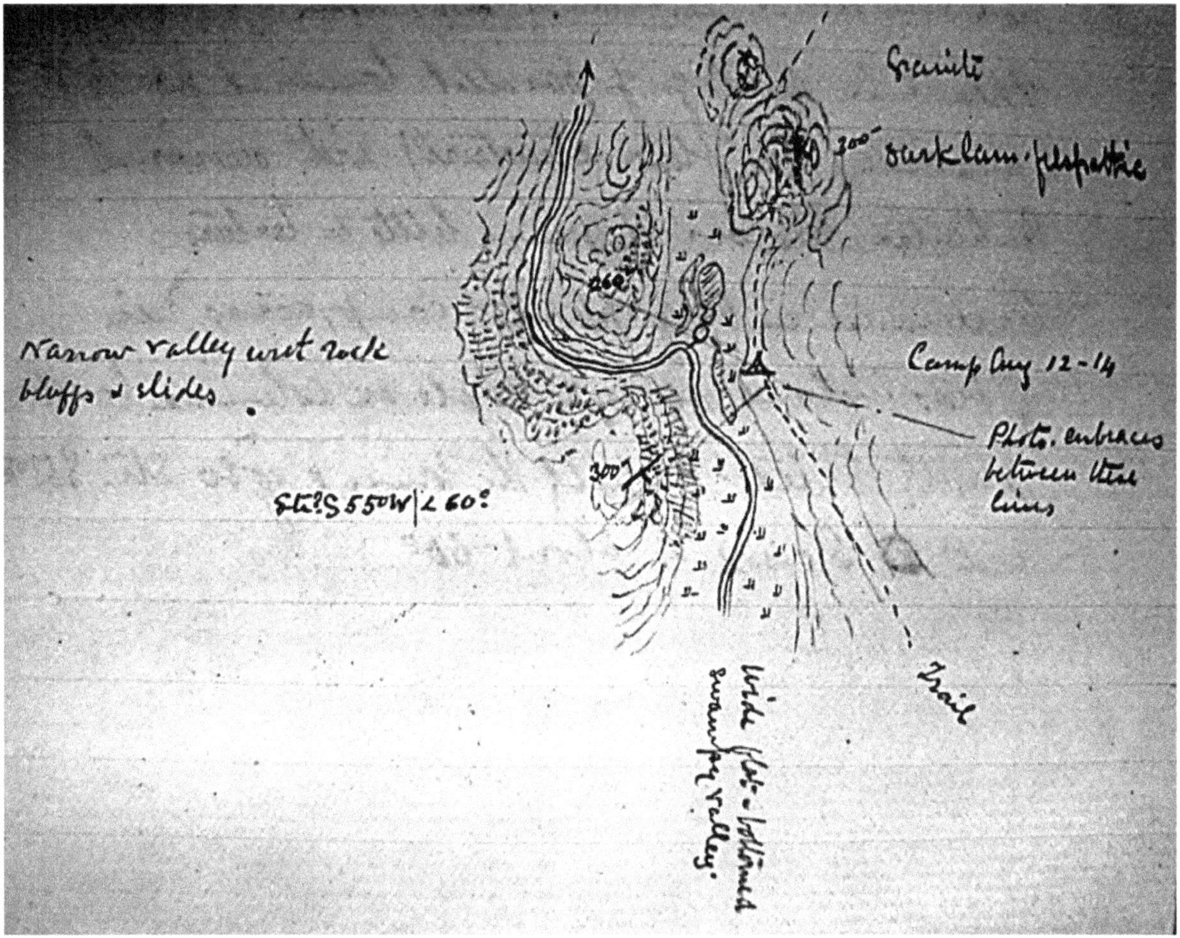

Dawson 1876 Geological Survey of Canada field notebooks.

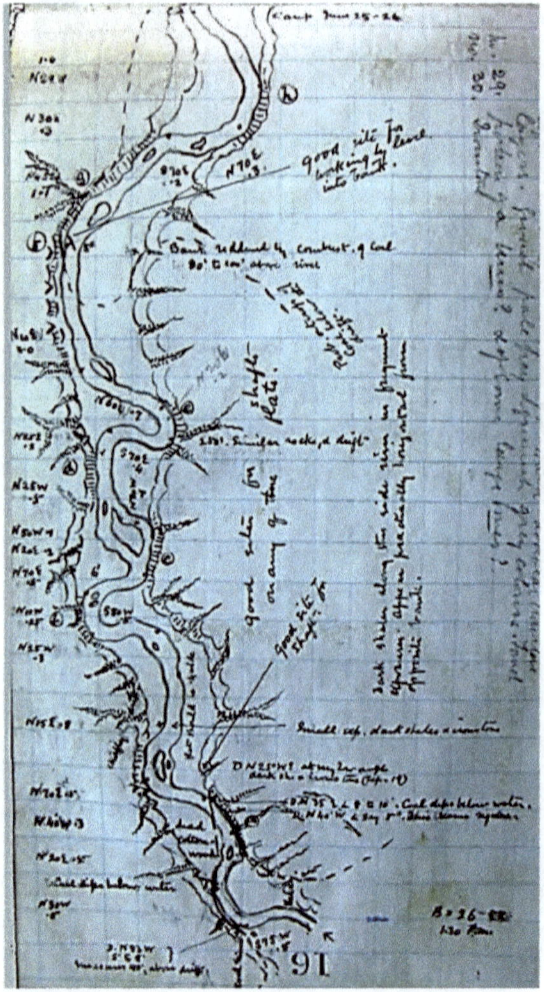

Dawson 1879 Geological Survey of Canada field notebooks.

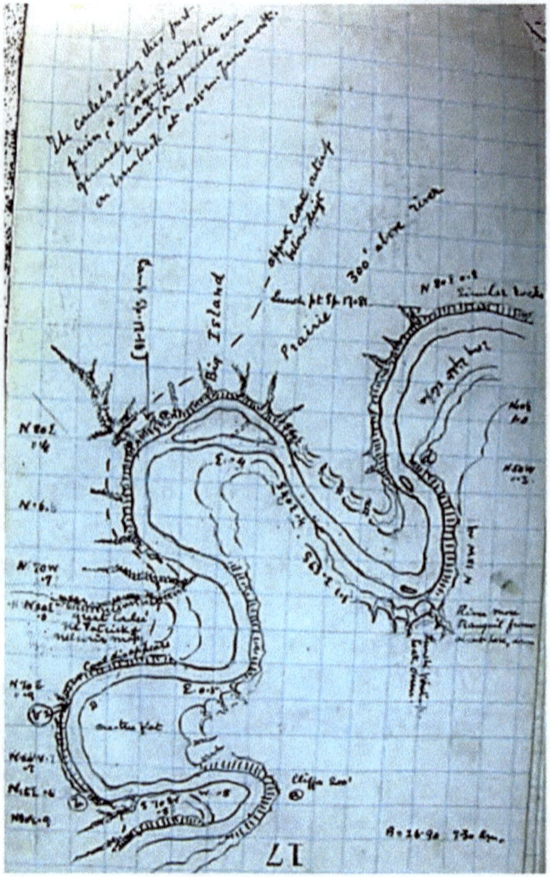

Dawson 1879 Geological Survey of Canada field notebooks.

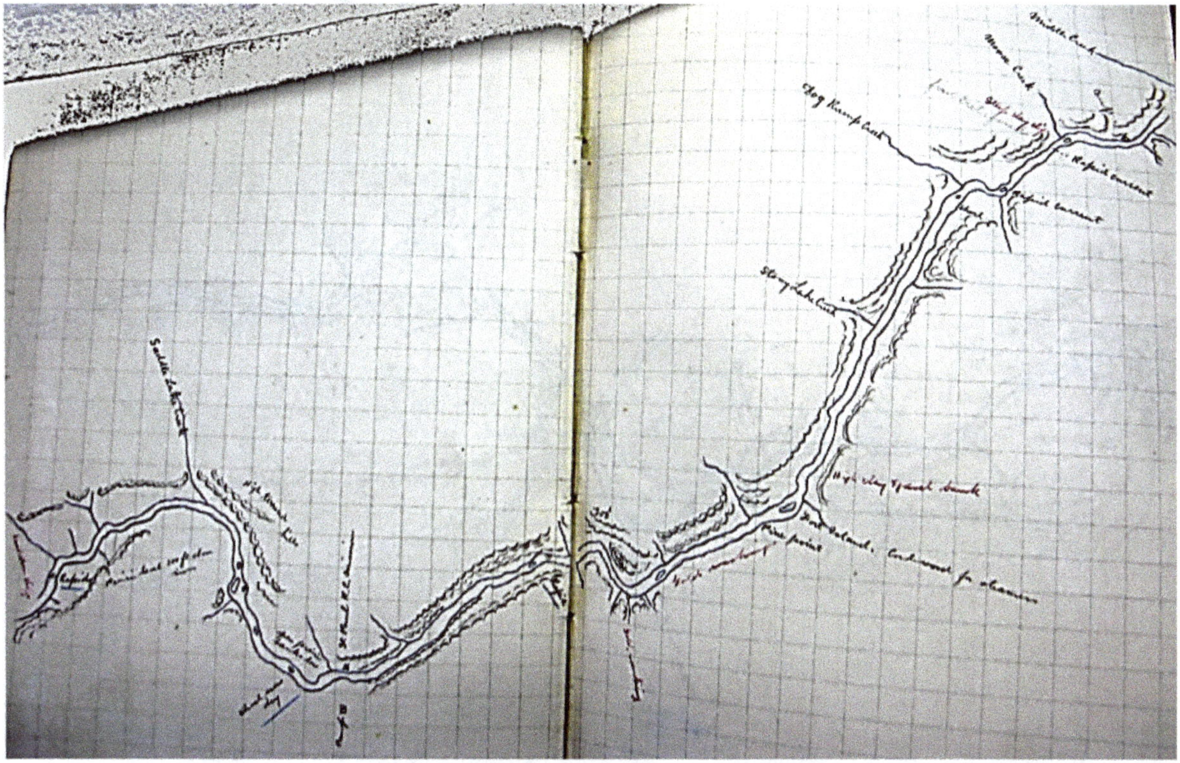

Dawson 1881 no 5. Geological Survey of Canada field notebooks.

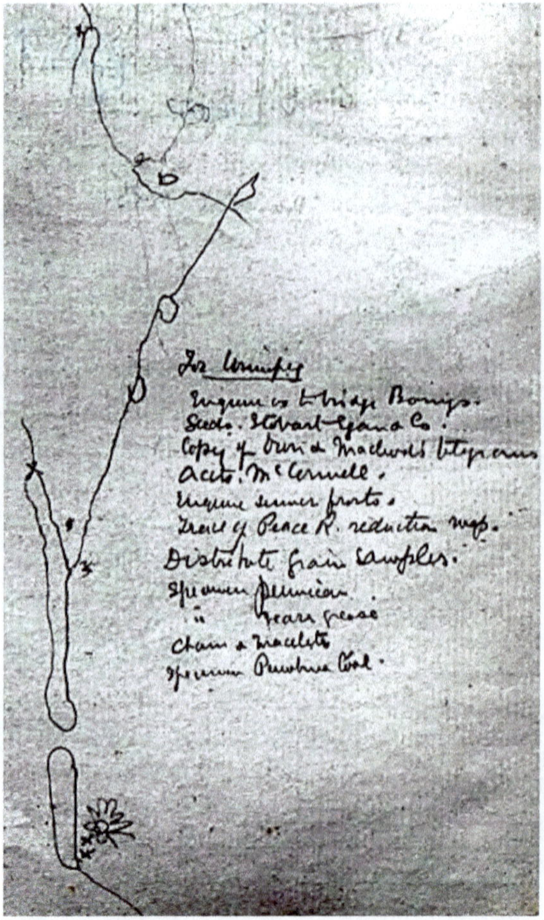

Dawson 1879. McGill niversity
1022 6 135 248

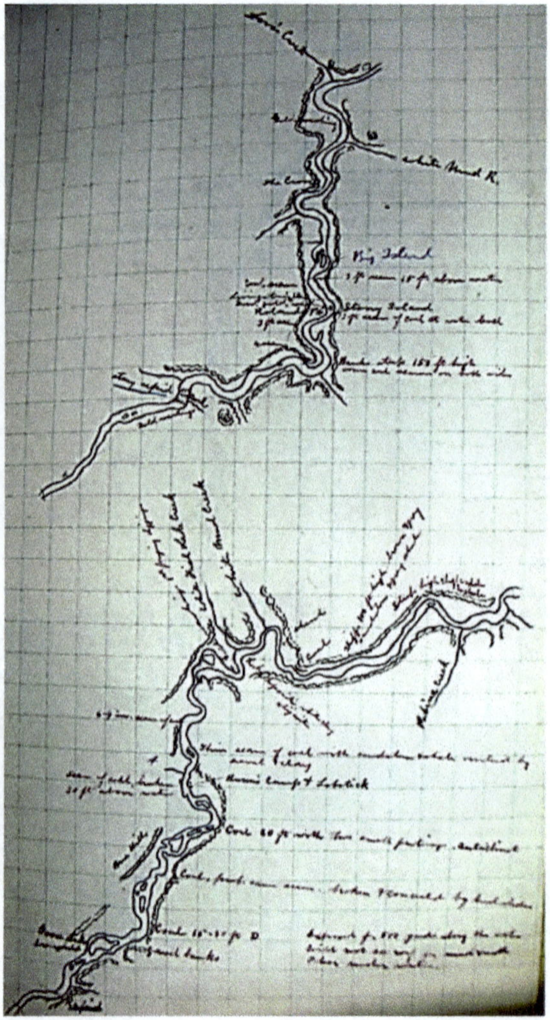

Dawson 1881 no 5. Geological Survey of Canada field notebooks.

Dawson 1898. Geological Survey of Canada field notebooks.

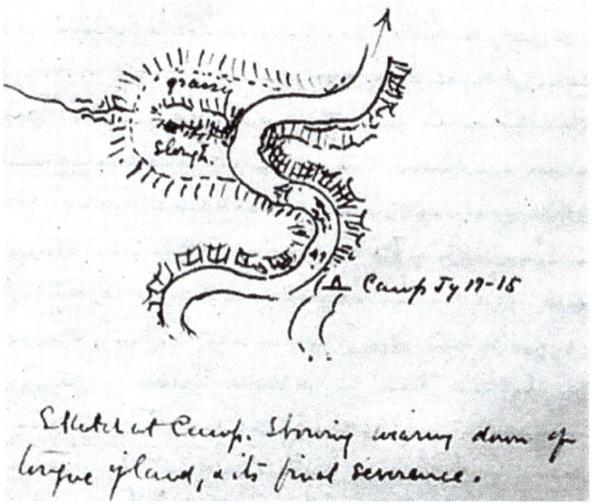

Dawson 1881. McGill niversity 1022 6 137 98

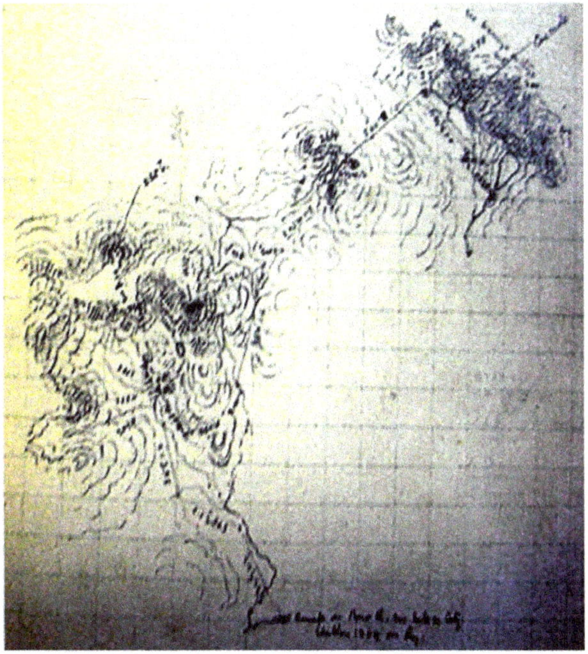

Dawson 1884. Geological Survey of Canada field notebooks.

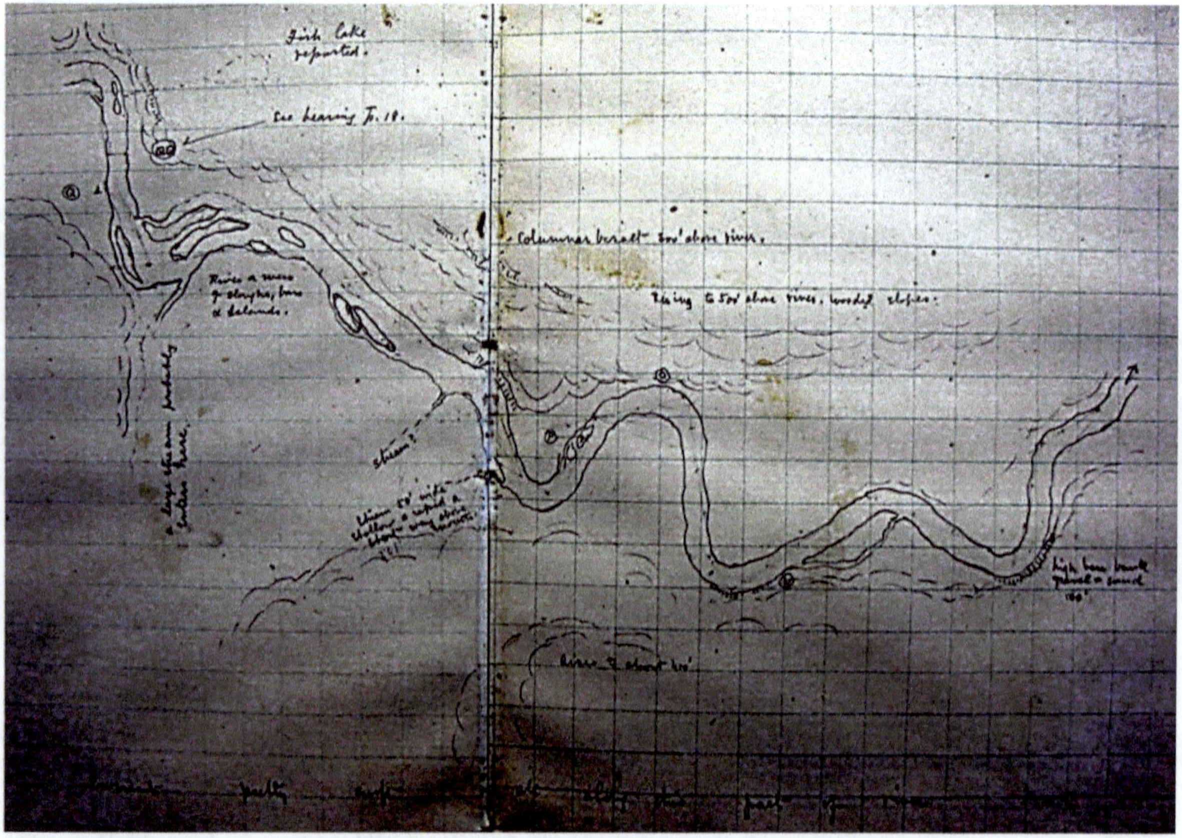

Dawson 1887. Geological Survey of Canada field notebooks.

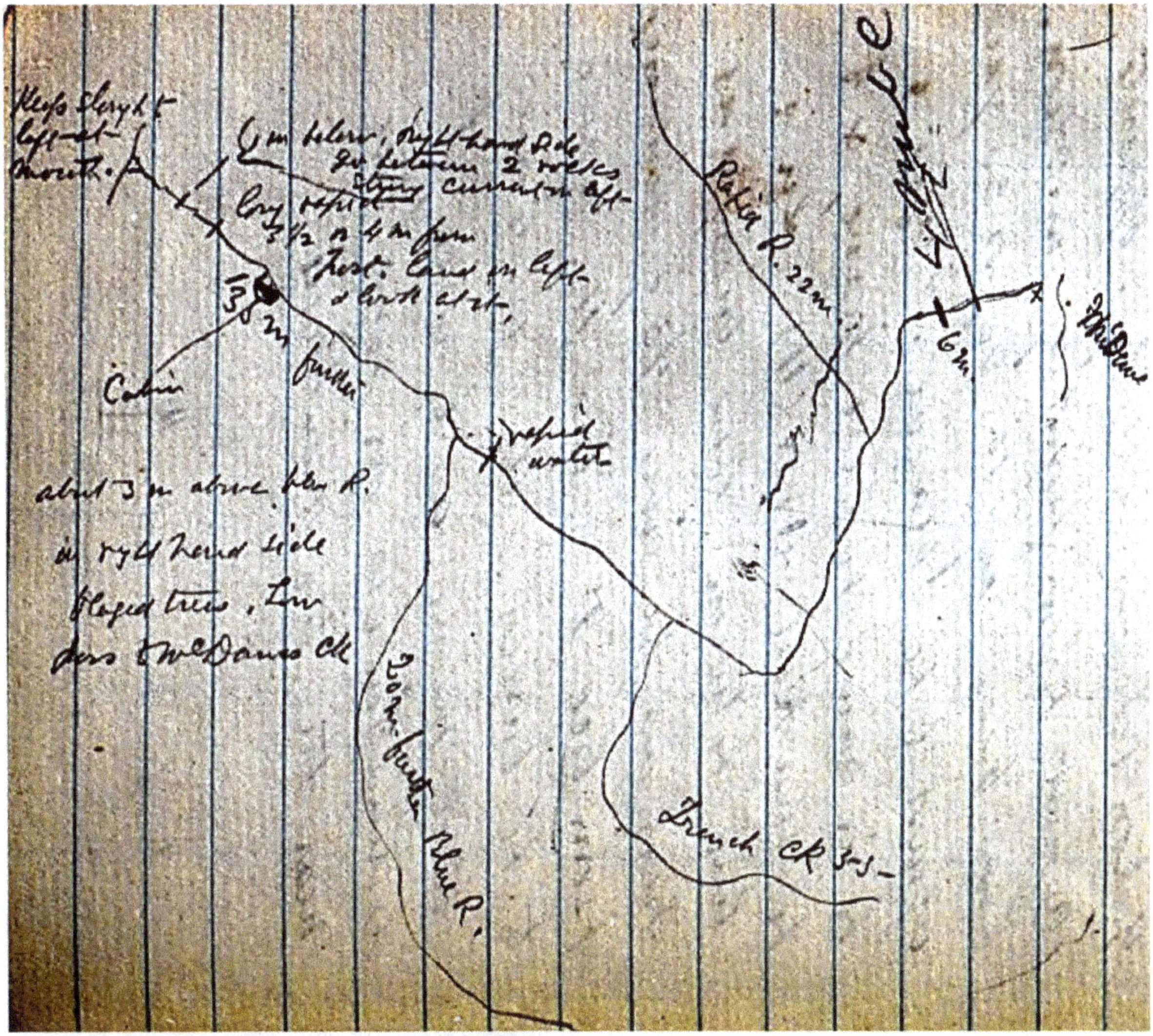

Dawson 1887. McGill niversity 1022 6 144 186

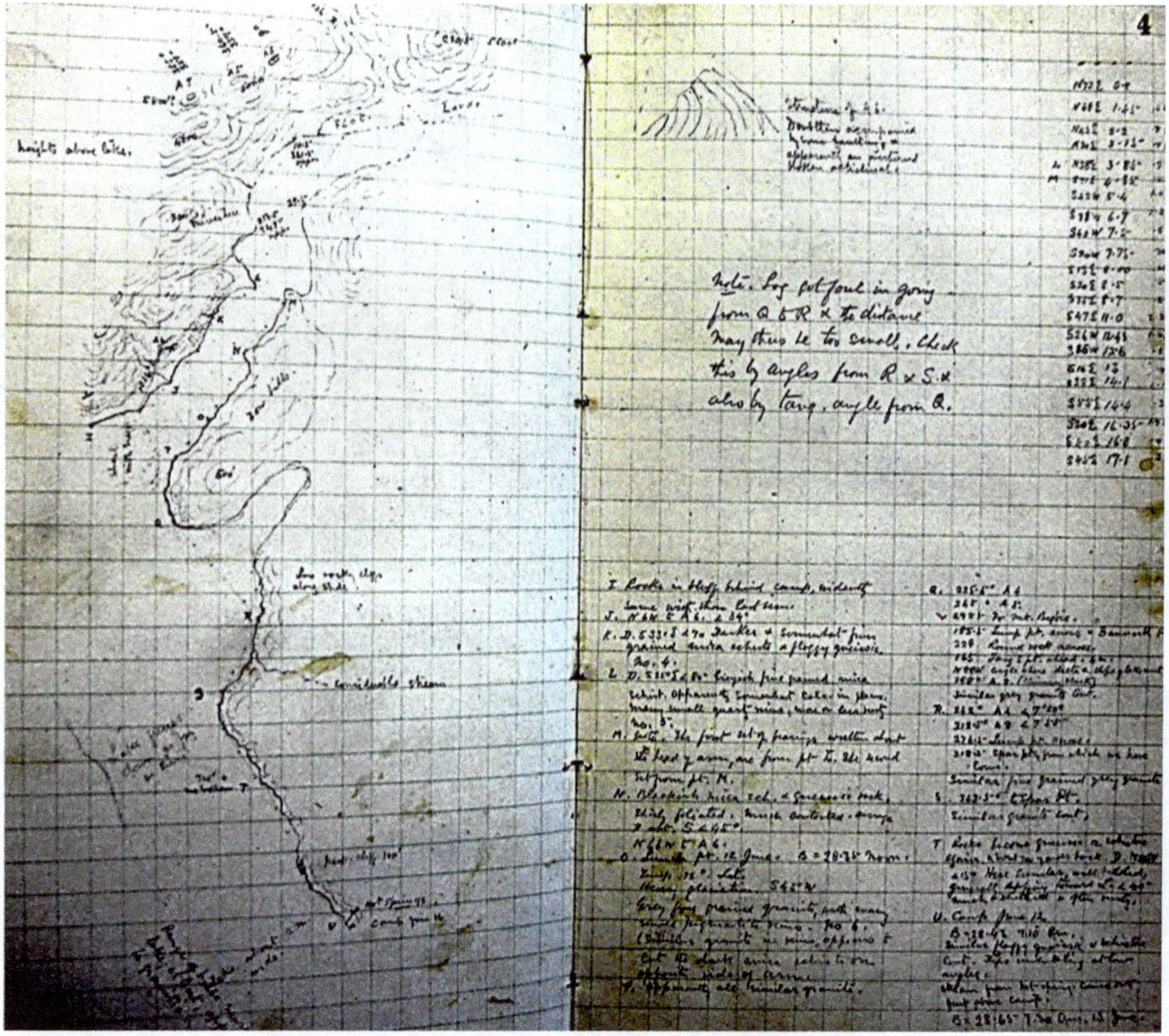

Dawson 1888 no 4. Geological Survey of Canada field notebooks.

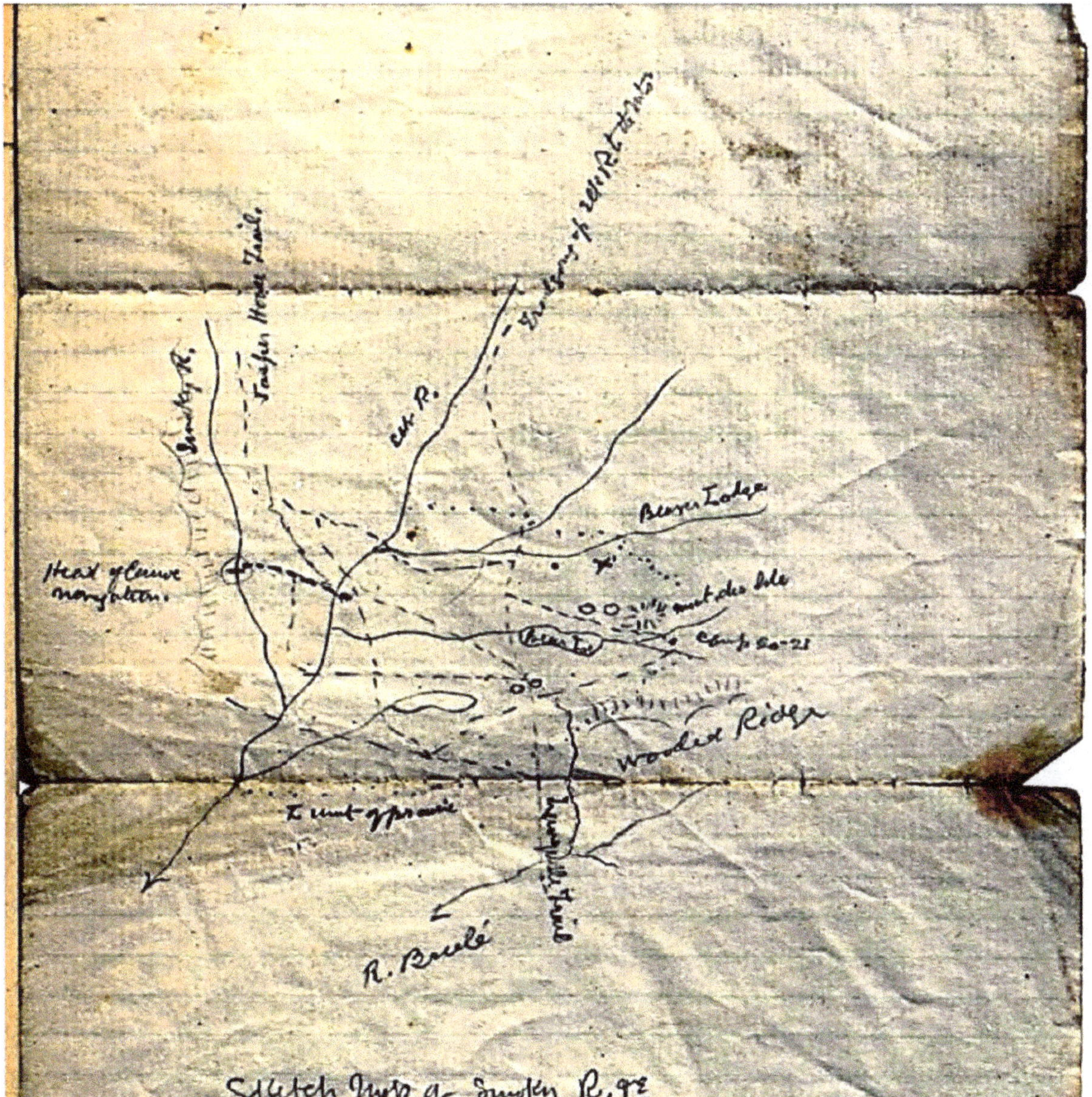

Dawson. McGill niversity 1022 6   45

Indigenous Peoples

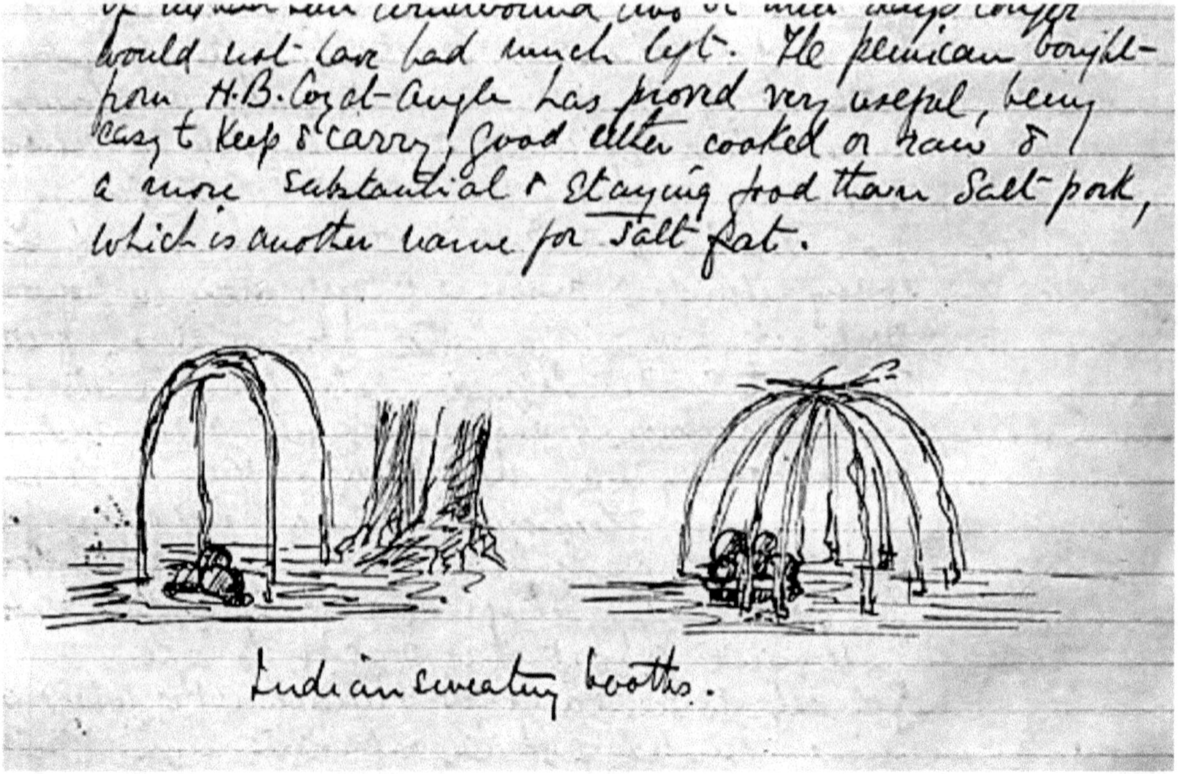

Dawson 1873 McGill niversity MG1022 6 178 46

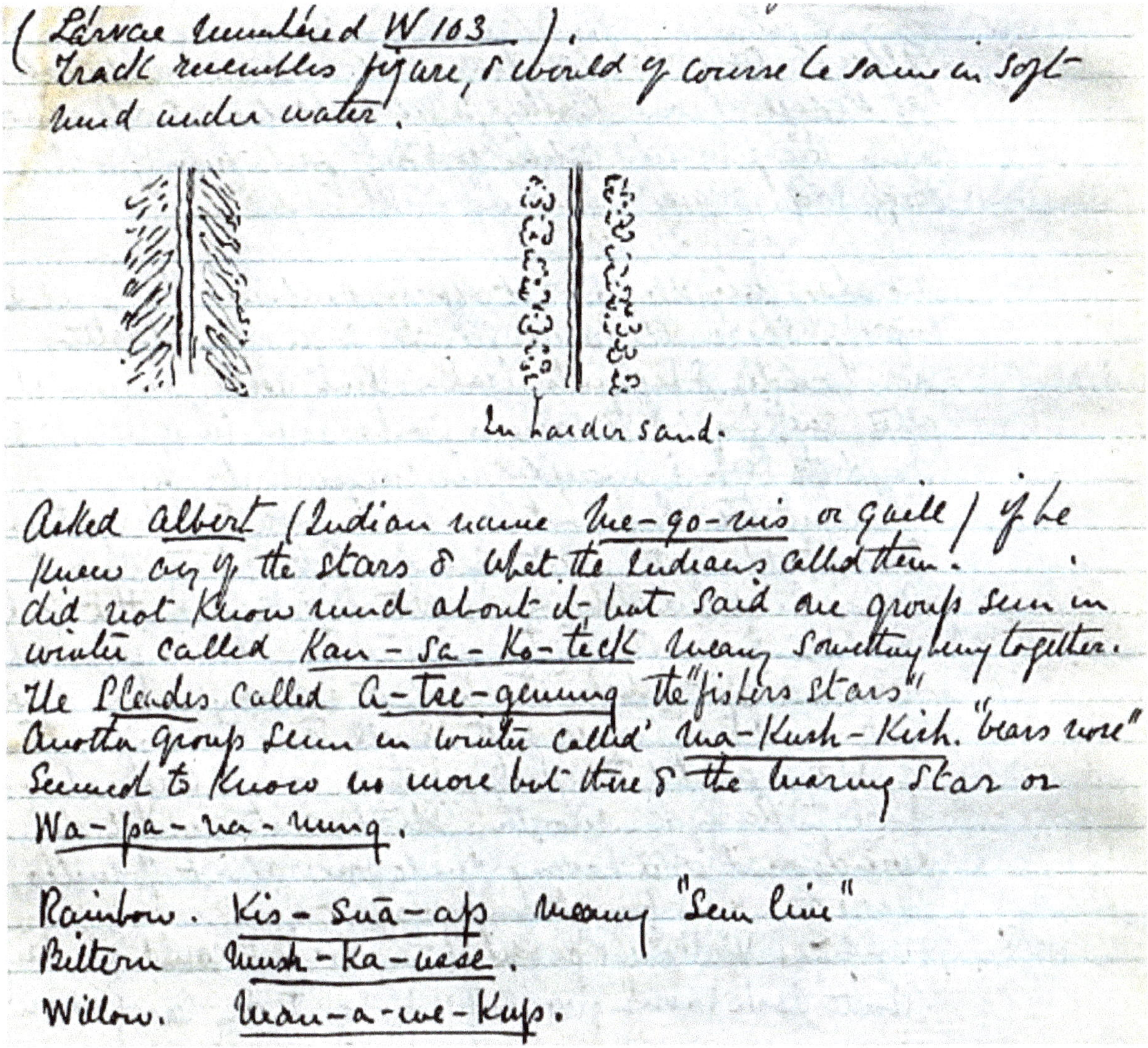

(Larvae numbered W 103).
Track resembles figure, & would of course be same in soft mud under water.

In harder sand.

Asked Albert (Indian name Ne-go-nis or gaile) if he knew any of the stars & what the Indians called them. did not know much about it, but said one group seen in winter called Kau-sa-ko-teck meaning something being together. The Pleiades called A-tee-gening the "fishers stars" Another group seen in winter called Ma-kush-kish. "bears nose" Seemed to know no more but these & the morning star or Wa-pa-na-nung.

Rainbow. Kis-suā-ap meaning "seen line"
Bittern Mush-ka-nese.
Willow. Nan-a-we-kups.

Dawson 1873 McGill niversity MG1022 6 178 37

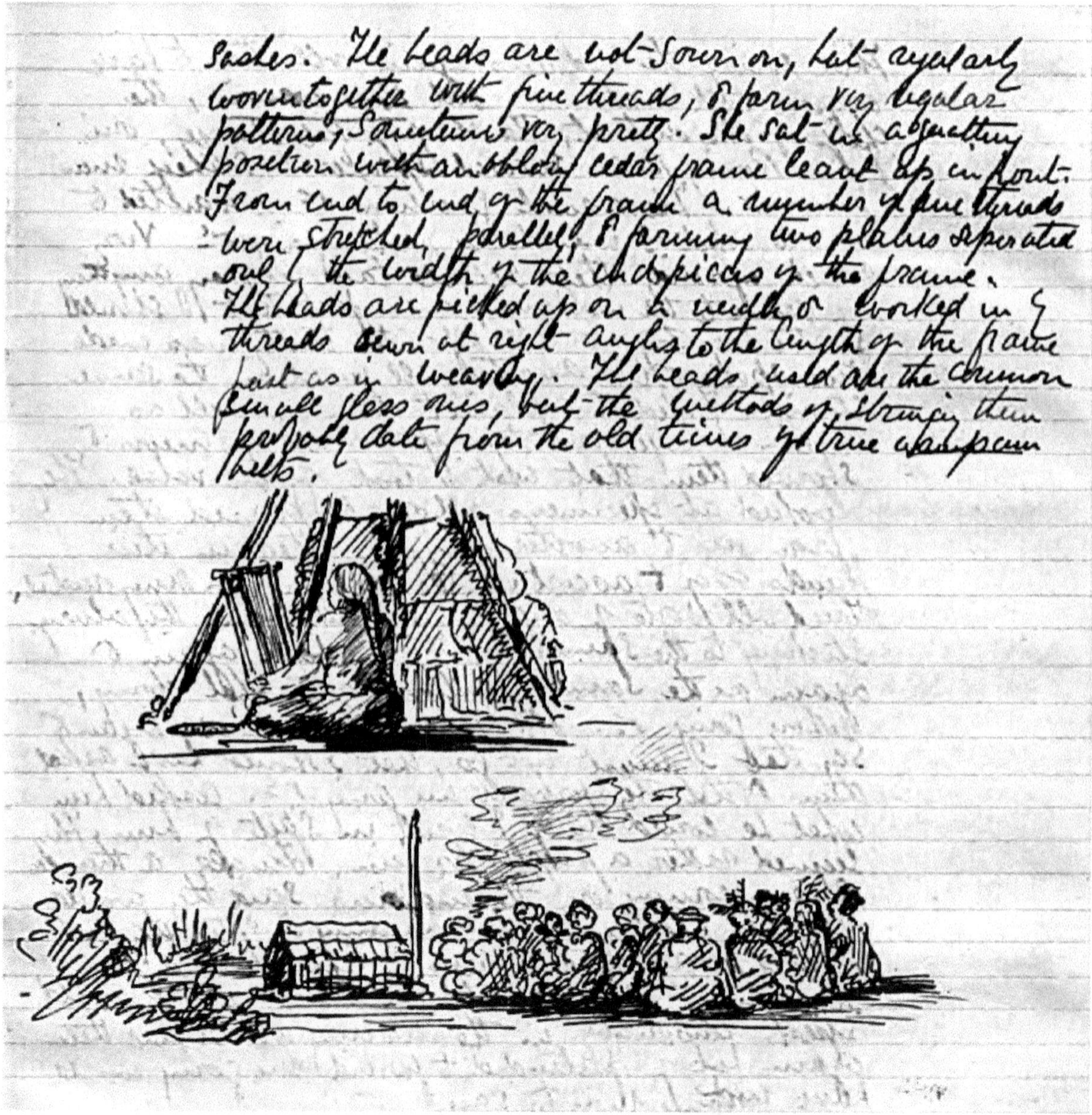

Sashes. The beads are not sewn on, but regularly woven together with fine threads, & form very regular patterns, sometimes very pretty. She sat in a squatting position with an oblong cedar frame leant up in front. From end to end of the frame a number of fine threads were stretched parallel, & forming two planes separated only by the width of the endpieces of the frame. The beads are picked up on a needle & worked in by threads sewn at right angles to the length of the frame just as in weaving. The beads used are the common small glass ones, but the methods of stringing them probably date from the old times of true wampum belts.

Dawson 1873 McGill niversity MG1022 6 178 57

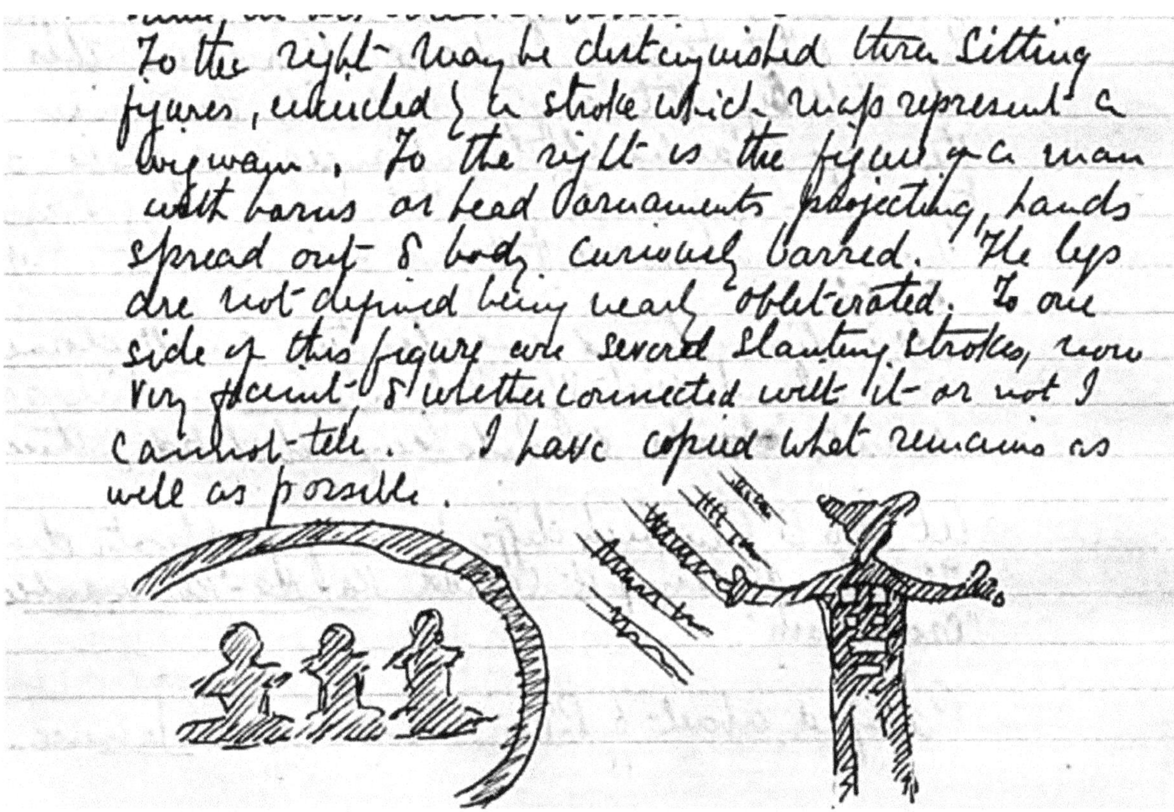

To the right may be distinguished three sitting figures, encircled by a stroke which may represent a wigwam. To the right is the figure of a man with horns or head ornaments projecting, hands spread out & body curiously barred. The legs are not defined being nearly obliterated. To one side of this figure are several slanting strokes, now very faint & whether connected with it or not I cannot tell. I have copied what remains as well as possible.

Dawson 1873 McGill niversity MG1022 6 178 63

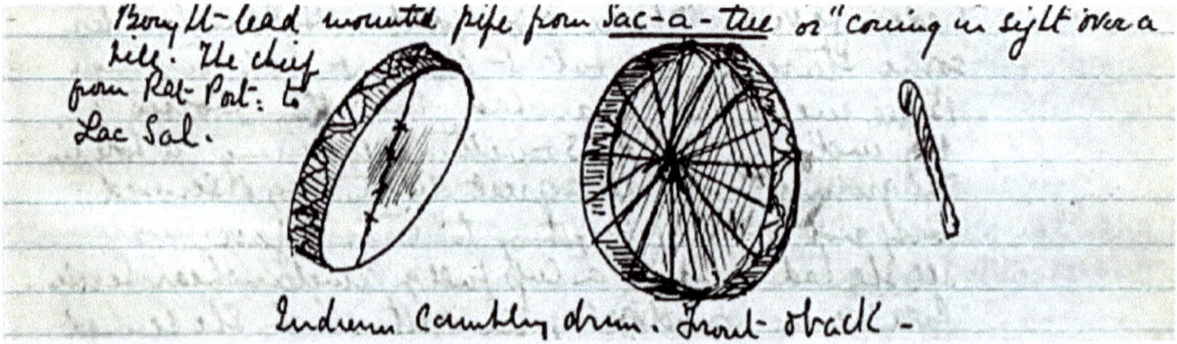
Bought lead mounted pipe from Sac-a-tee or "coming in sight over a hill." The chief from Rat-Port: to Lac Sal.

Indian Gambling drum. Front & back -

Dawson 1873 McGill niversity MG1022 6 178 61

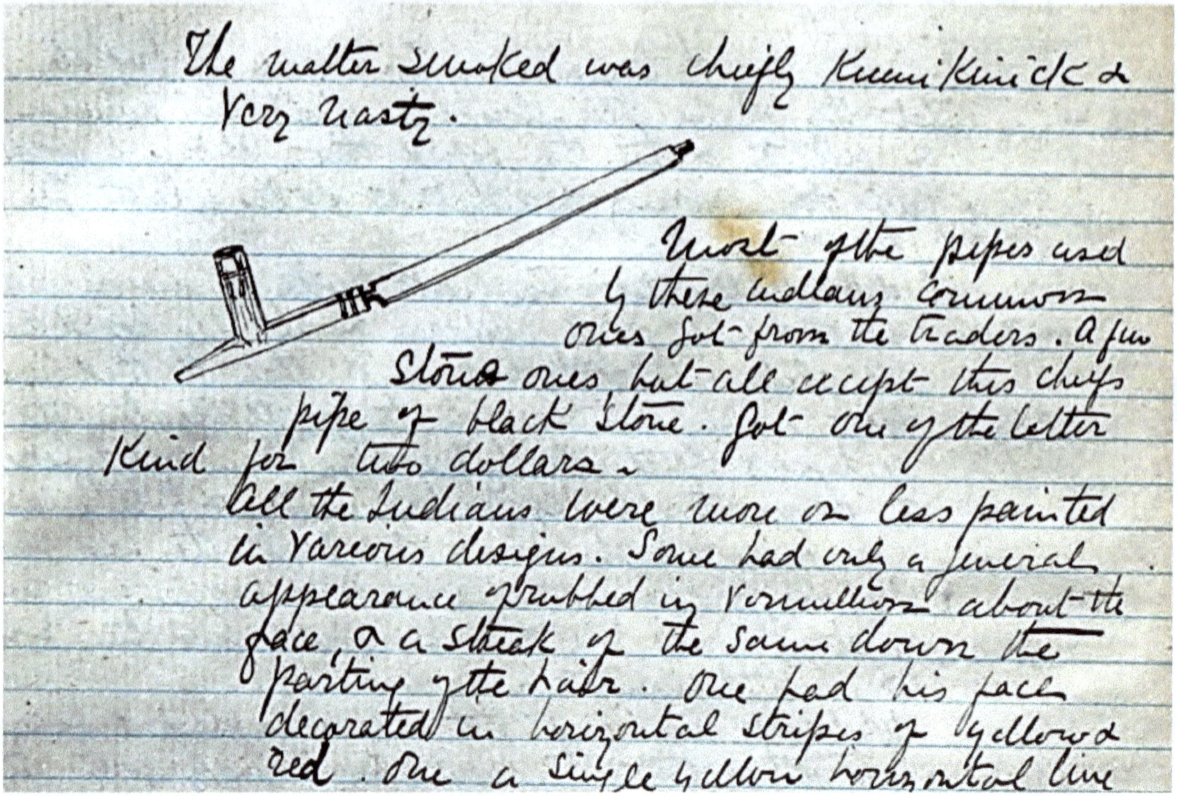
The matter smoked was chiefly Kinnikinick & very nasty.

Most of the pipes used by these Indians common ones got from the traders. A few stone ones but all except this chiefs pipe of black stone. Got one of the latter kind for two dollars.
All the Indians were more or less painted in various designs. Some had only a general appearance of rubbed in vermillion about the face, or a streak of the same down the parting of the hair. One had his face decorated in horizontal stripes of yellow & red. One a single yellow horizontal line

Dawson 1874 McGill niversity MG1022 6 180 270

5 or 6 Carts. No 1. driven by materfamilias. No 2. by Sons wife, or eldest daughter. The rest very generally by children which frequently almost babies. Lodge poles tied alongside Carts & projecting behind. A cloud of skirmishers consisting of dogs & loose ponies, the former not unfrequently hauling travailles & cooking [illegible] wonderfully helpless & timorous when rushed out upon by a pack of dogs from our camp. Men mostly ugly. Long haired & rough looking, dressed in blue coats with brass buttons, or skin suits. Women hard looking & soon wrinkled, wearing calico dresses of dark colours & with a turban of similar material wound round the head.

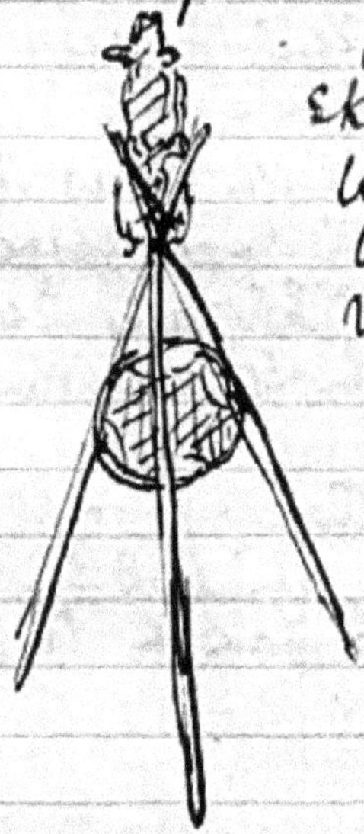

Camp. W. Fork 6 am.
B. 26.24
T. 41° WSW. clear.

Camp at Lake 5 m. E of E. Fork 8 AM
B. 26.21
T. 42° N.W. light - clear.

Wind rose with Sun blew strong westerly all day & fell at sundown.

Dawson 1874 McGill niversity MG1022 6 180 237

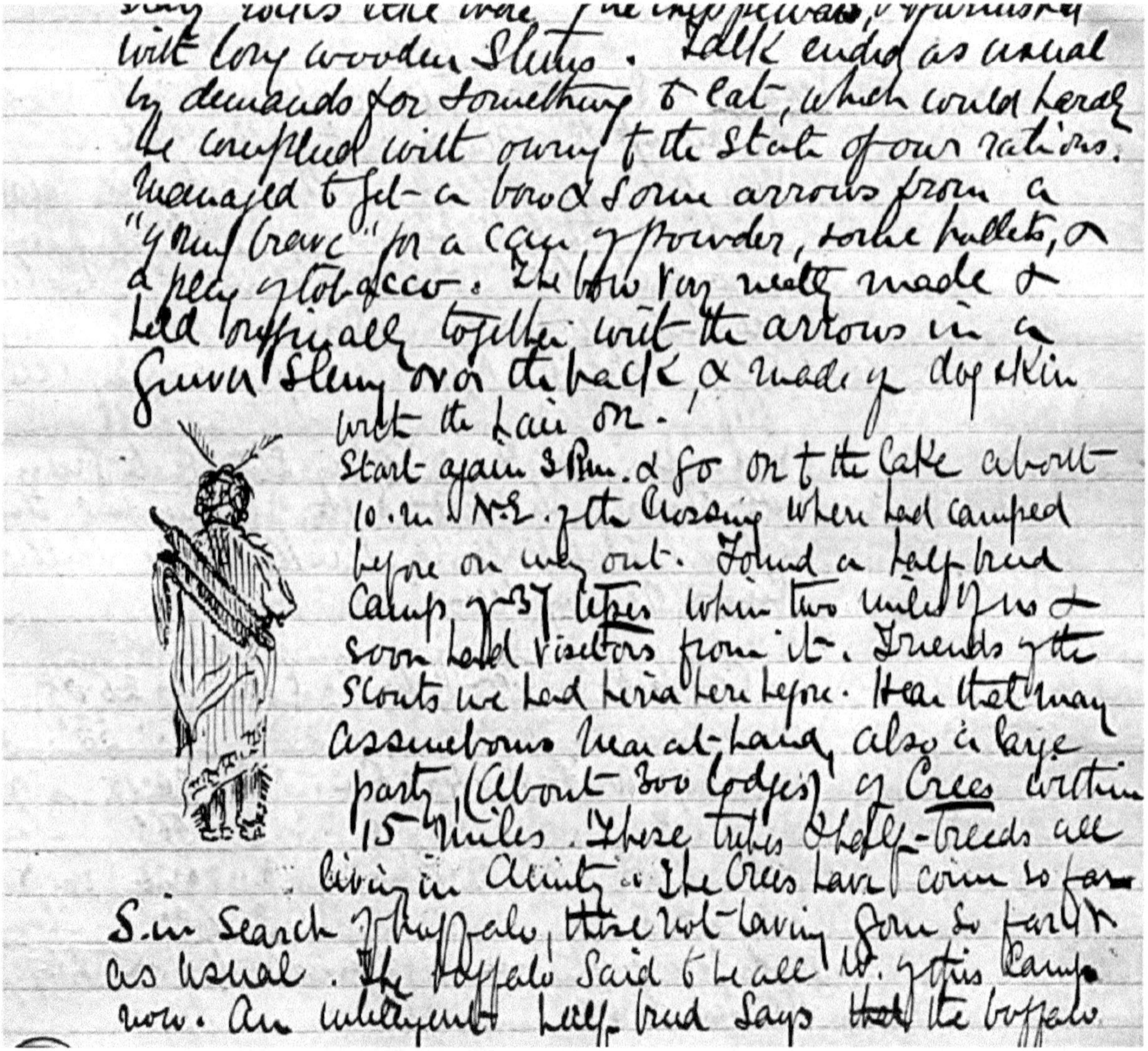

... the Chippewas, ... with long wooden Stems. Talk ended as usual by demands for something to eat, which could hardly be complied with owing to the state of our rations. Managed to get a bow & some arrows from a "young brave" for a can of powder, some bullets, & a plug of tobacco. The bow very neatly made & held [illegible] together with the arrows in a quiver slung over the back, & made of dog skin with the hair on. Start again 3 P.m. & go on to the Lake about 10 m. N.E. of the Crossing where had camped before on way out. Found a half breed camp of 37 tepees when two miles of us & soon had visitors from it. Friends of the Scouts we had seen here before. Hear that many Assiniboines near at hand, also a large party (about 300 lodges) of Crees within 15 miles. These tribes & half-breeds all living in amity. The Crees have come so far S. in search of buffalo, these not having gone so far N as usual. The buffalo said to be all W. of this camp now. An intelligent half breed says the buffalo

Dawson 1874 McGill niversity MG1022 6 180 233

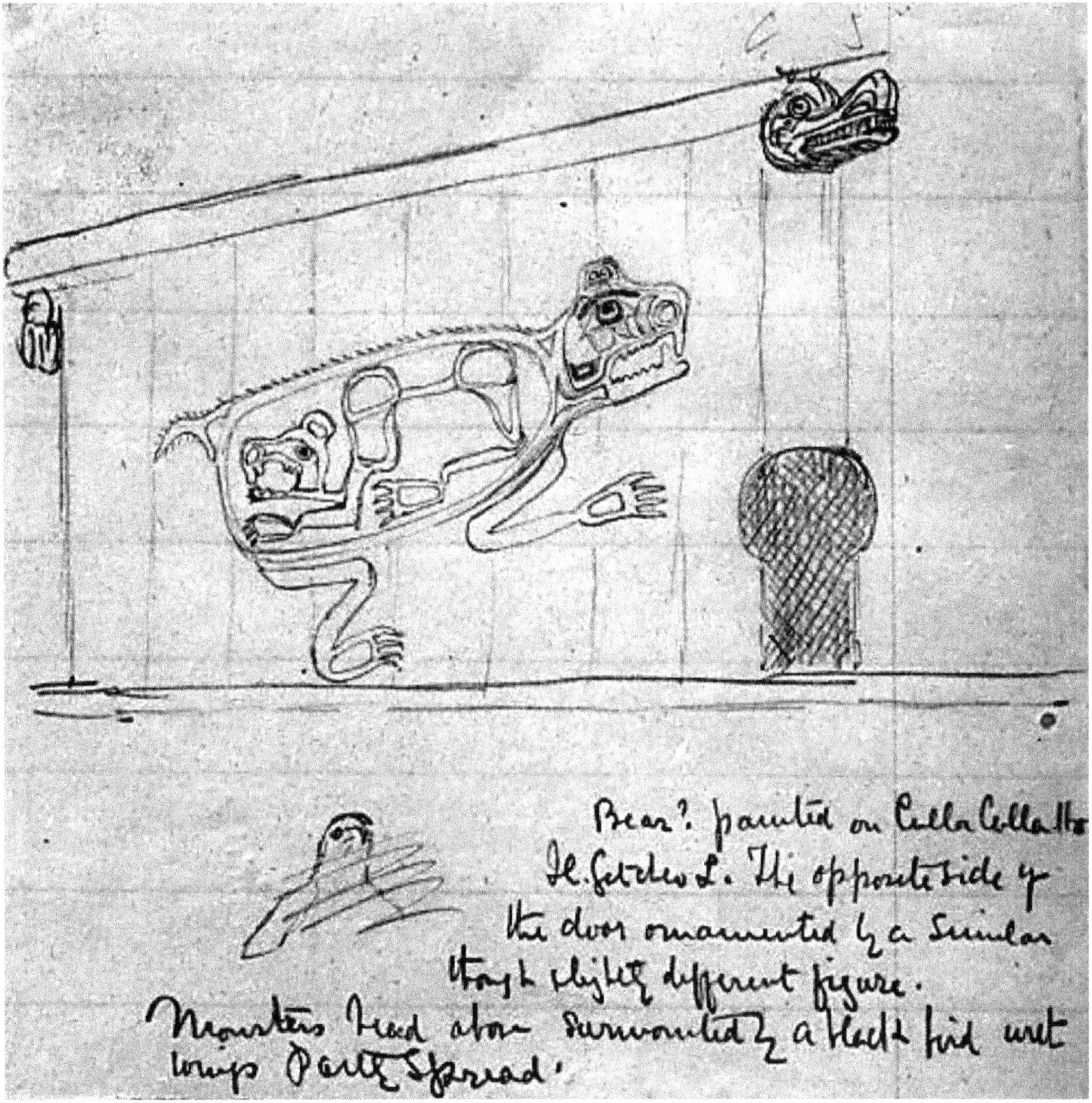

Dawson 1875 McGill niversity MG1022 6 182 116

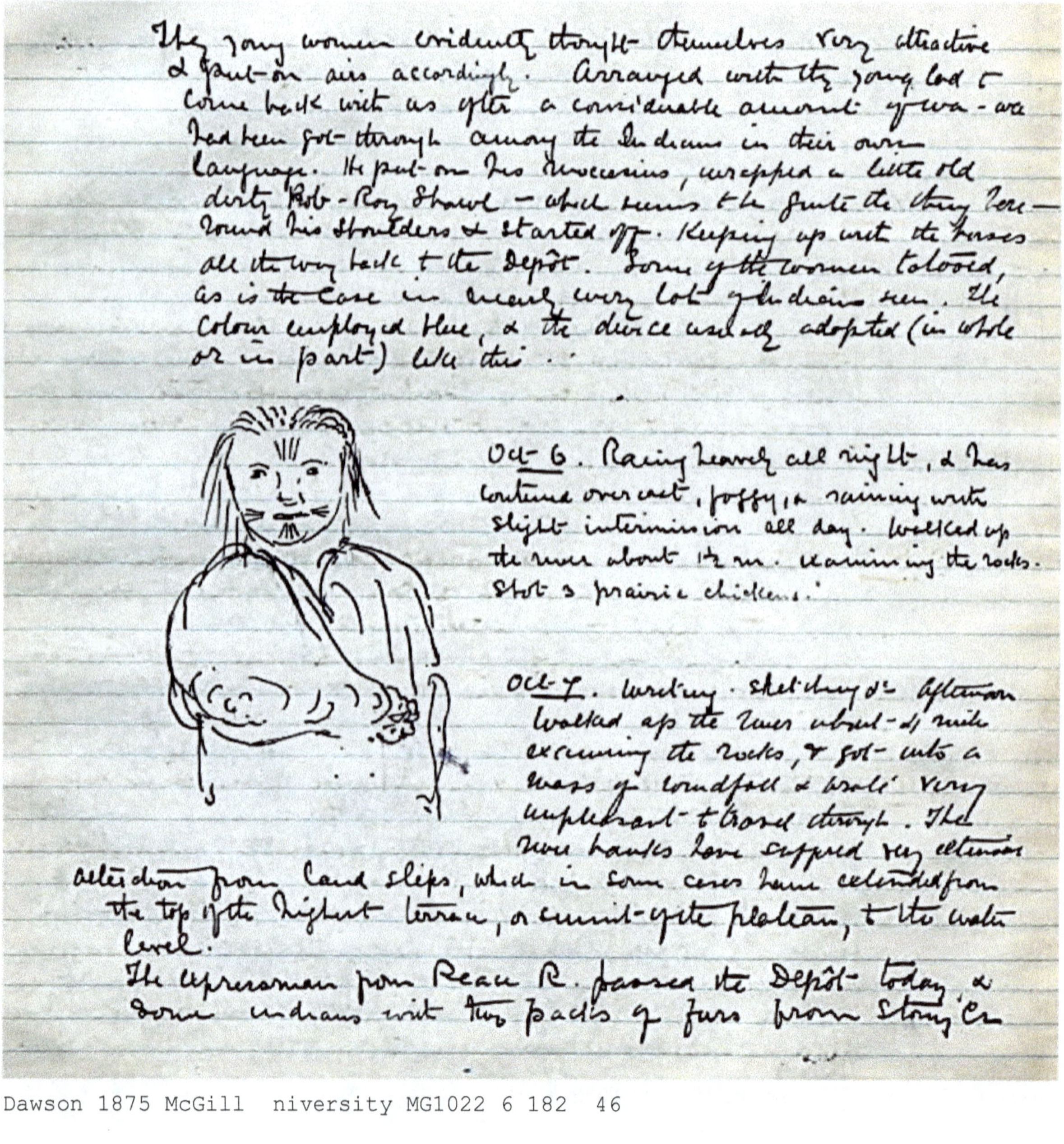

The young women evidently thought themselves very attractive & put on airs accordingly. Arranged with the young lad to come back with us after a considerable amount of wa-wa had been got through among the Indians in their own language. He put on his moccasins, wrapped a little old dirty Rob-Roy Shawl — which seems to be quite the thing here — round his shoulders & started off. Keeping up with the horses all the way back to the Depôt. Some of the women tatooed, as is the case in nearly every lot of Indians seen. The colour employed blue, & the device usually adopted (in whole or in part) like this:

Oct 6. Raining heavily all night, & has continued overcast, foggy, & raining with slight intermission all day. Walked up the river about 1½ m. examining the rocks. Shot 3 prairie chickens.

Oct 7. Writing. Sketching &c. Afternoon walked up the river about 4 miles examining the rocks, & got into a mass of windfall & brulé very unpleasant to travel through. The river banks have suffered very extensive alteration from land slips, which in some cases have extended from the top of the highest terrace, or summit of the plateau, to the water level.

The expressman from Peace R. passed the Depôt today & some Indians with two packs of furs from Stony Cr

Dawson 1875 McGill niversity MG1022 6 182 46

Dawson 1876 McGill niversity MG1022 6 131 44

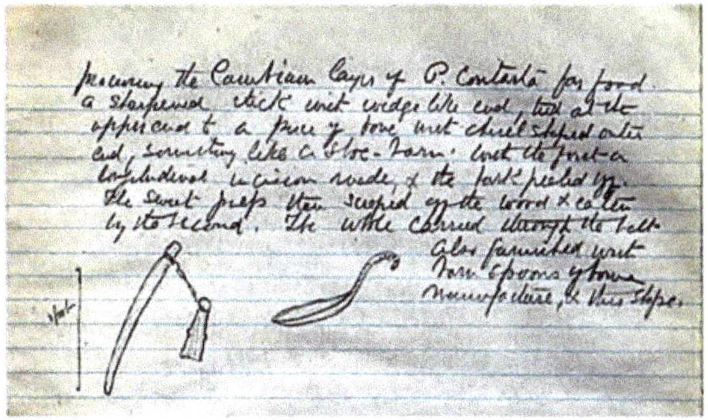

Dawson 1876 McGill niversity MG1022 6 131 23

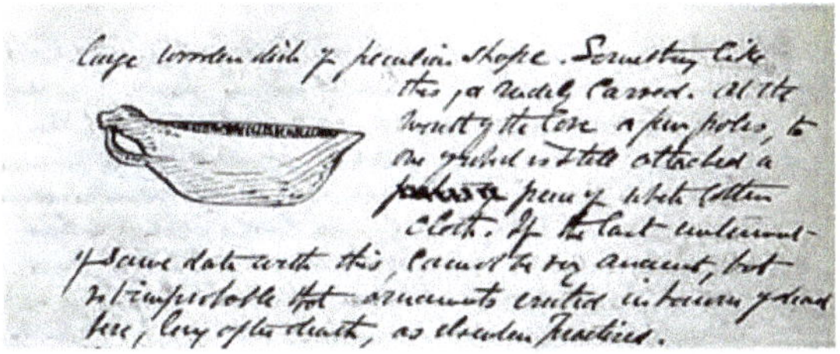

Dawson 1878 McGill niversity MG1022 6 134 198

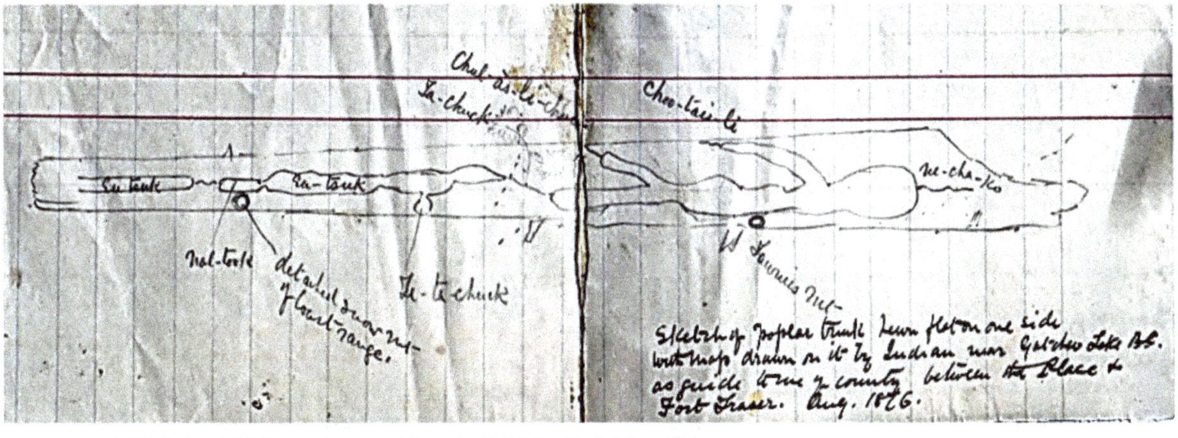

Dawson 1876 McGill niversity MG1022 6 131 58

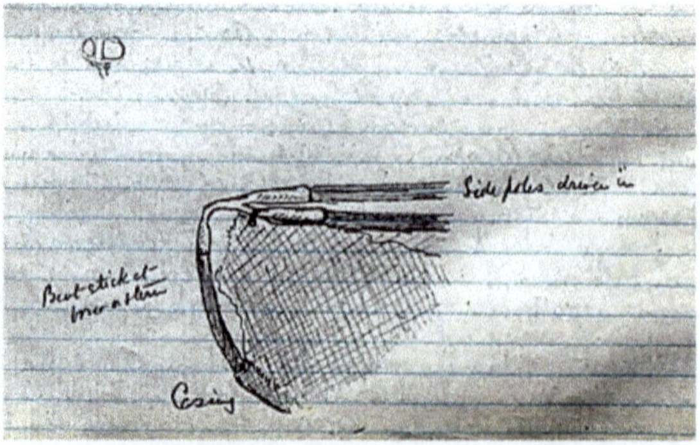

Dawson 1876 McGill niversity
MG1022 6 131 184

Dawson 1881 McGill niversity
MG1022 6 137 166

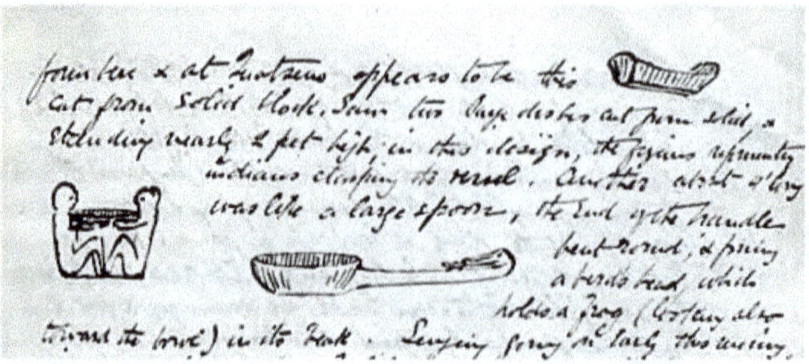

Dawson 1878 McGill niversity MG1022 6
134 215

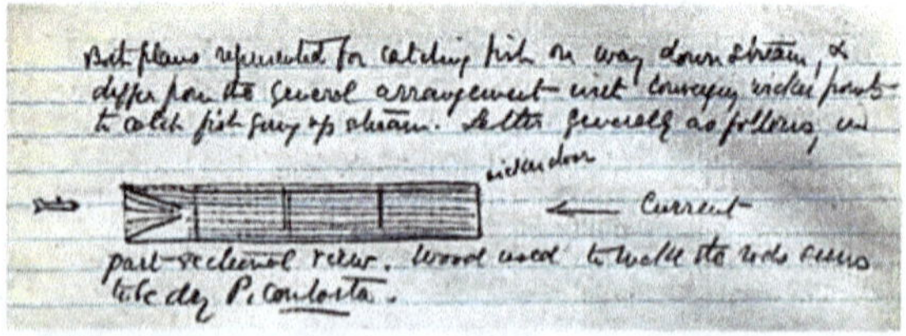

Dawson 1876 McGill niversity
MG1022 6 131 43

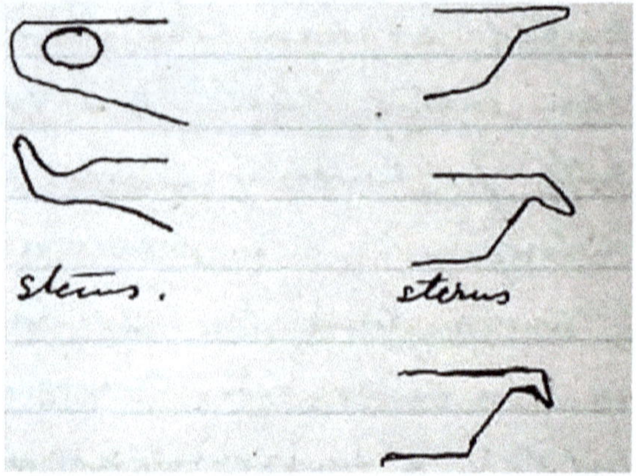

Dawson 1891 McGill niversity
MG1022 6 64 46

Dawson 1881 no 4. Geological Survey of Canada field notebooks.

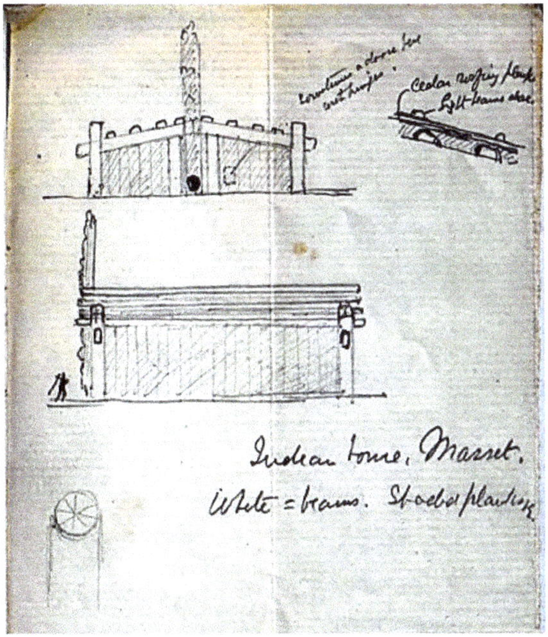

Dawson 1878 McGill niversity MG1022 6 134 5

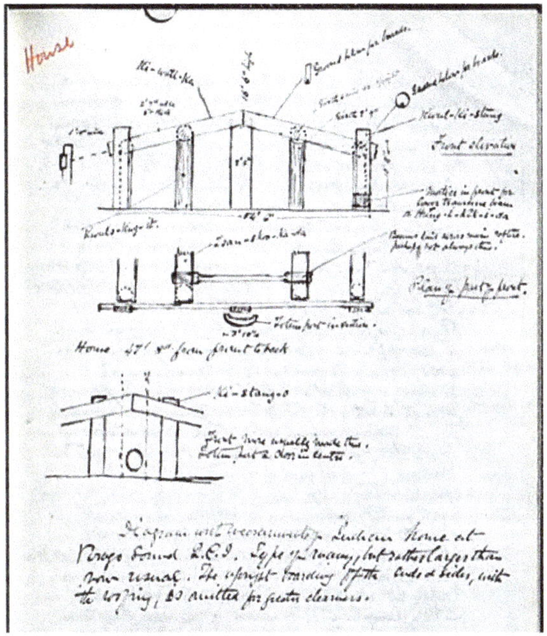

Dawson 1878 McGill niversity MG1022 6 134 157

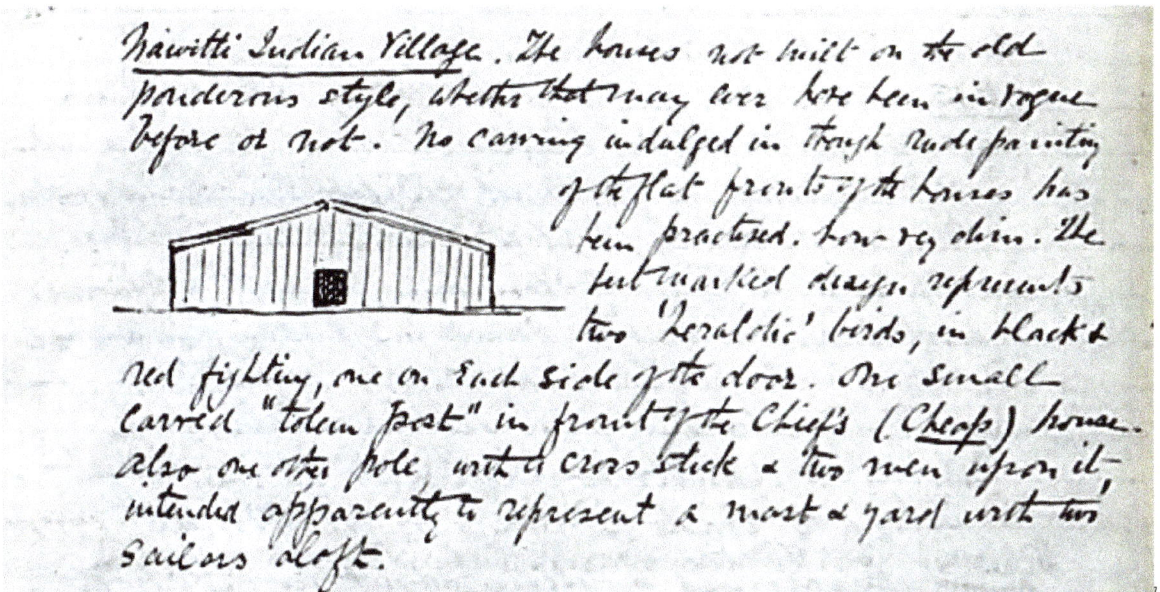
Naiwitti Indian Village. The houses not built on the old ponderous style, whether that may ever have been in vogue before or not. No carving indulged in though rude painting of the flat fronts of the houses has been practised. Low – very clean. The first marked diagr. represents two 'heraldic' birds, in black & red fighting, one on each side of the door. One small carved "totem post" in front of the Chief's (Cheap) house. Also one other pole with a cross stick & two men upon it, intended apparently to represent a mast & yard with two sailors aloft.

Dawson 1878 McGill niversity MG1022 6 134 10

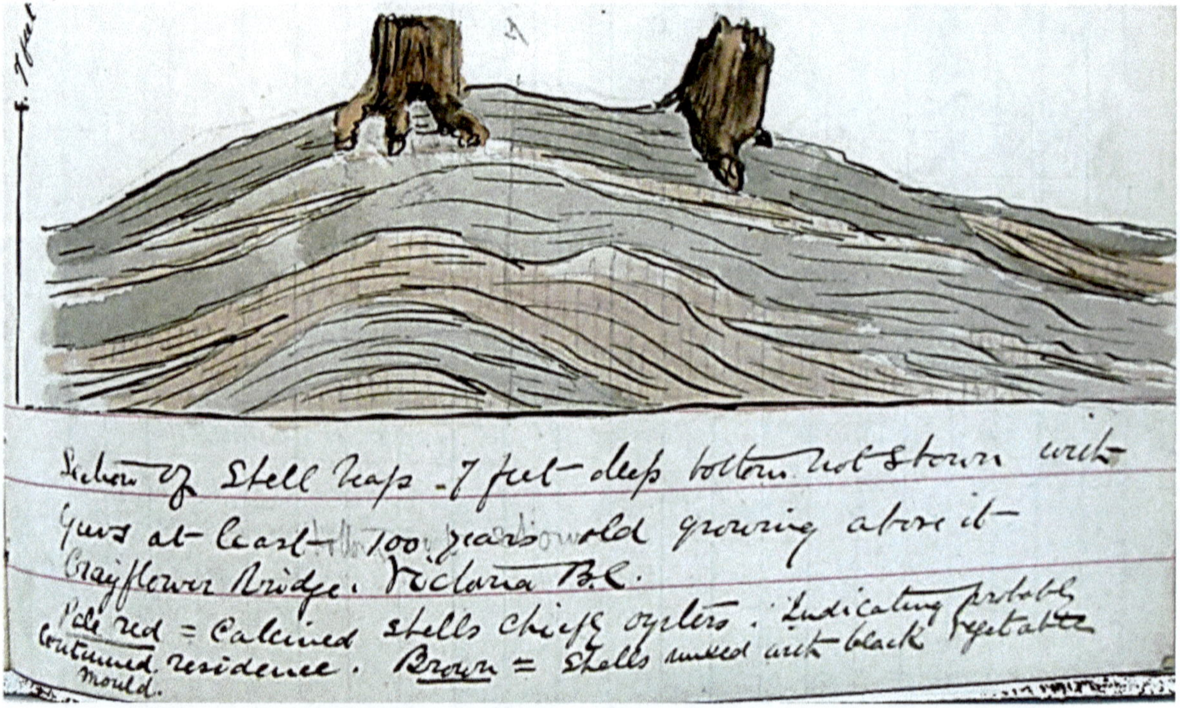

Dawson 1879 Geological Survey of Canada field notebooks. riginal in colour.

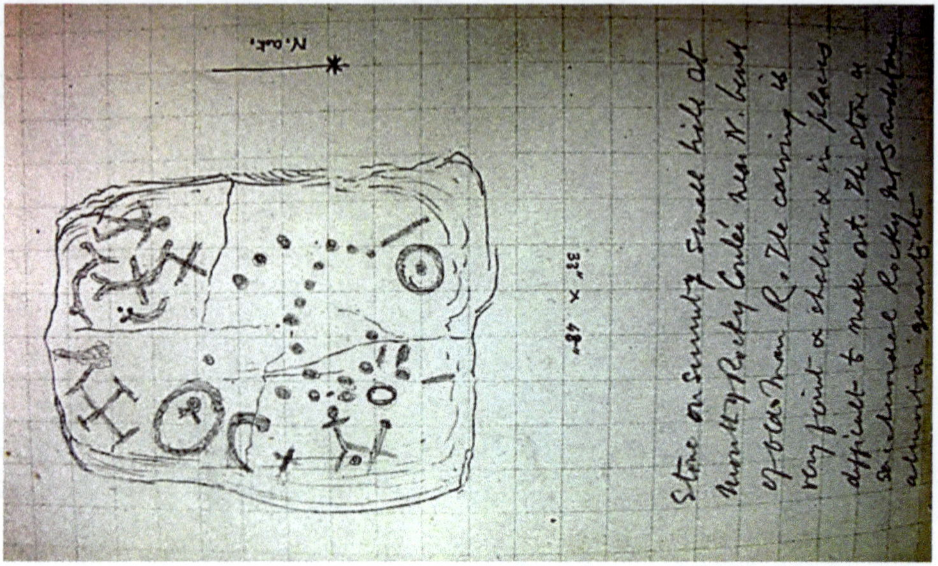

Dawson 1881 no 5. Geological Survey of Canada field notebooks.

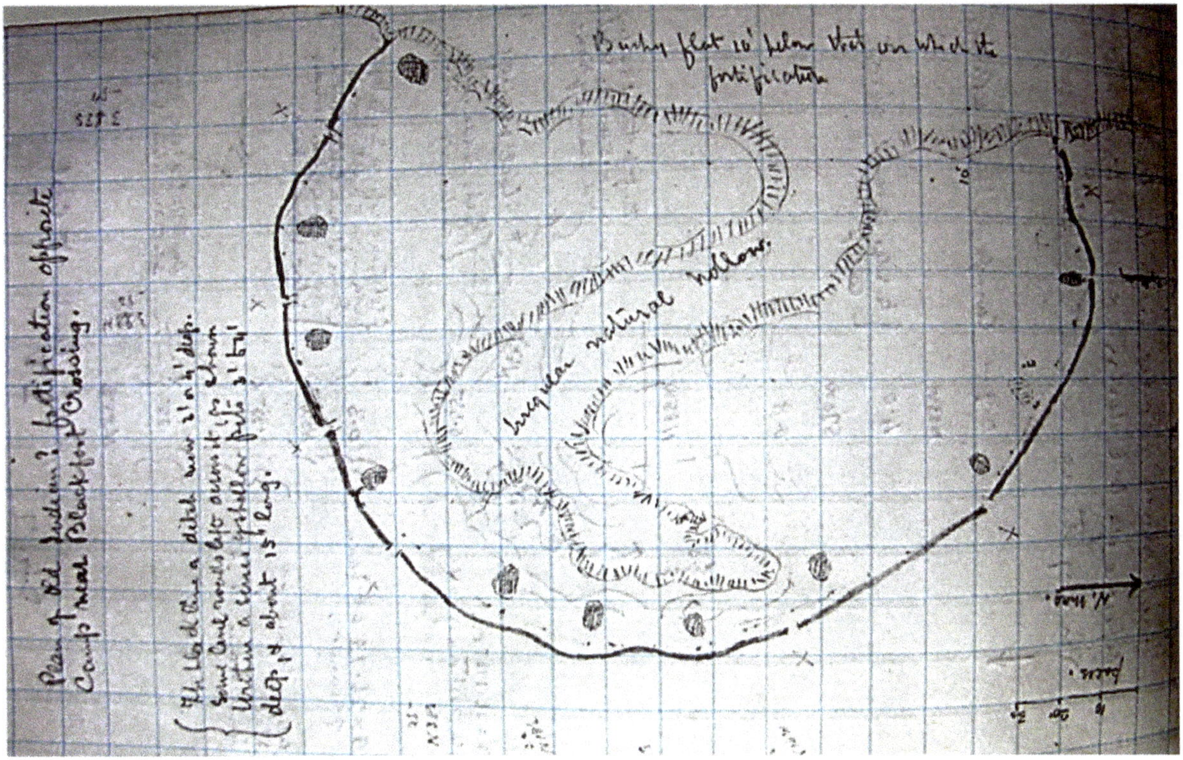

Dawson 1881 no3. Geological Survey of Canada field notebooks.

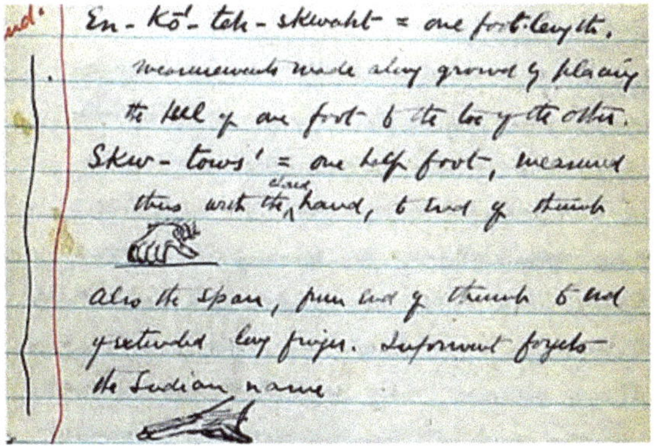

Dawson 1889 McGill niversity
MG1022 6 146 120

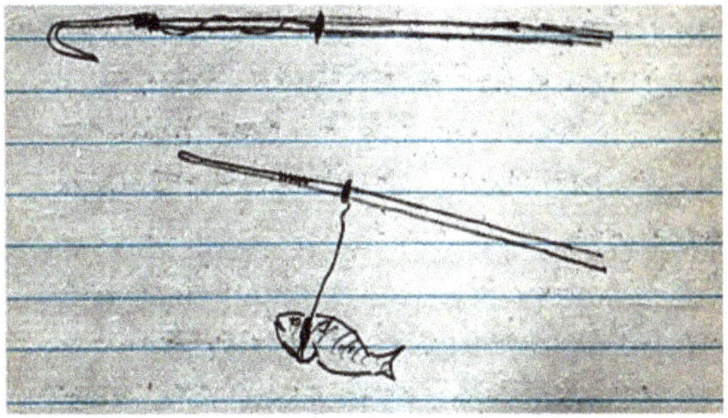

Dawson 1885 McGill niversity
MG1022 6 142 80

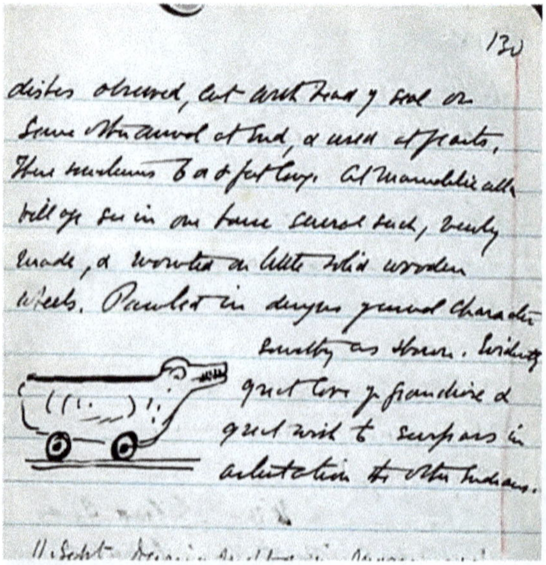

Dawson 1885 McGill niversity
MG1022 6 142 135

Dawson 1891 McGill niversity
MG1022 6 64 104

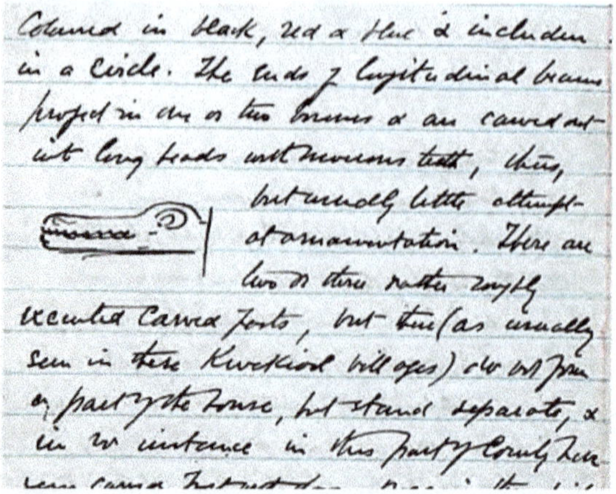

Dawson 1885 McGill niversity
MG1022 6 142 129

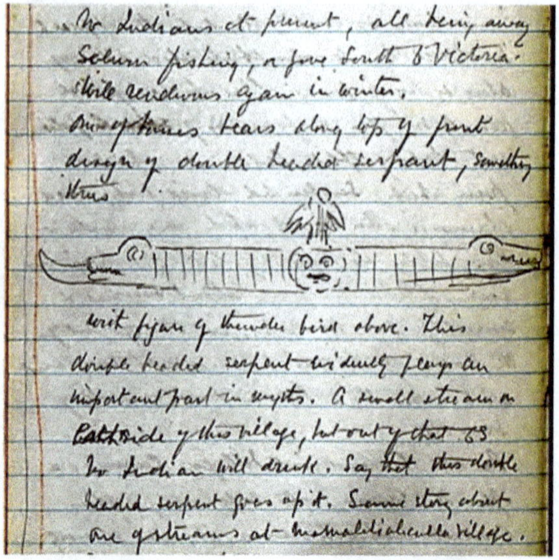

Dawson 1885 McGill niversity
MG1022 6 142 134

Dawson 1890 McGill niversity
MG1022 6 147 179

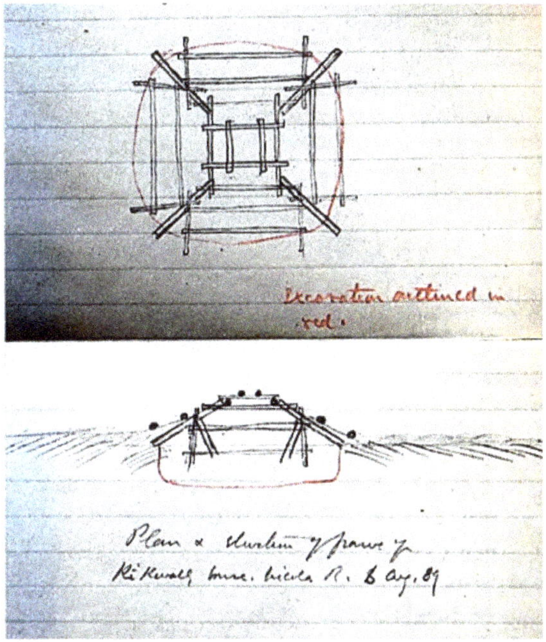

Dawson 1889 McGill niversity
MG1022 6 146 78

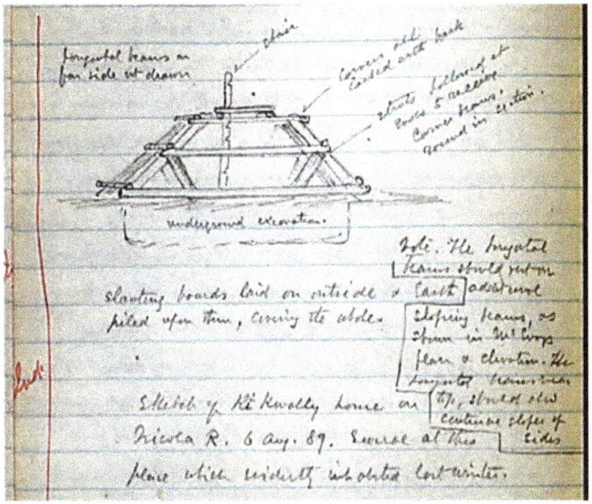

Dawson 1889 McGill niversity
MG1022 6 146 76

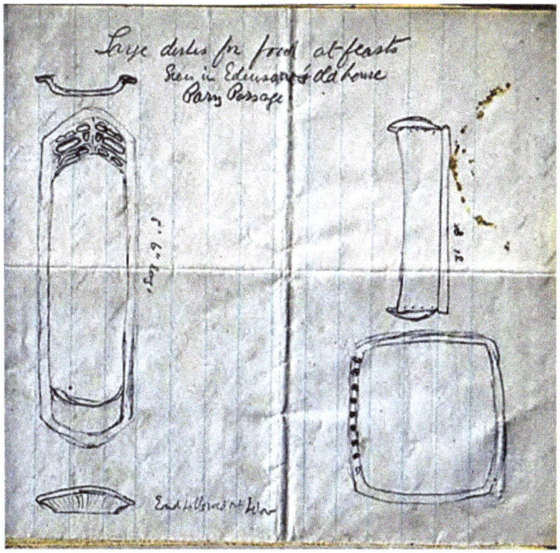

Dawson 1878 McGill niversity
MG1022 6 134 49 3

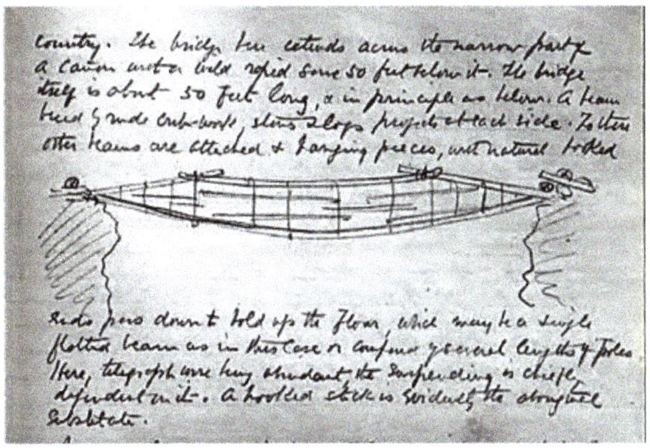

Dawson 1879 McGill niversity
MG1022 6 135 81

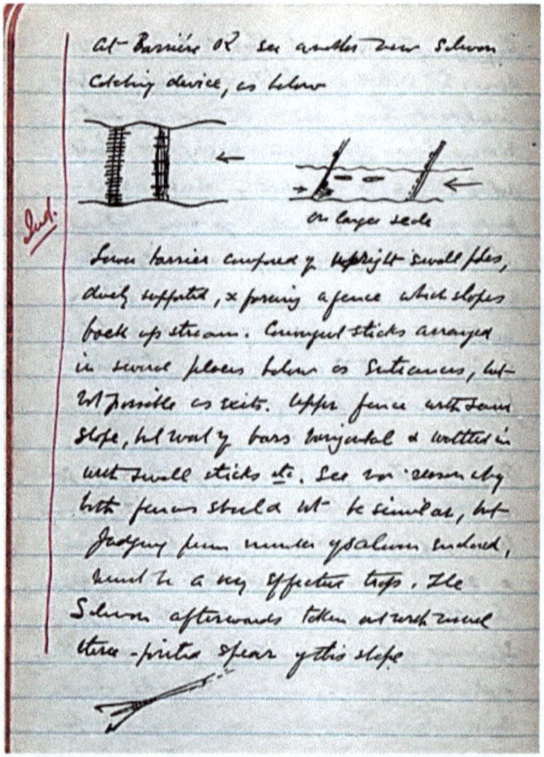

Dawson 1890 McGill niversity
MG1022 6 147 92

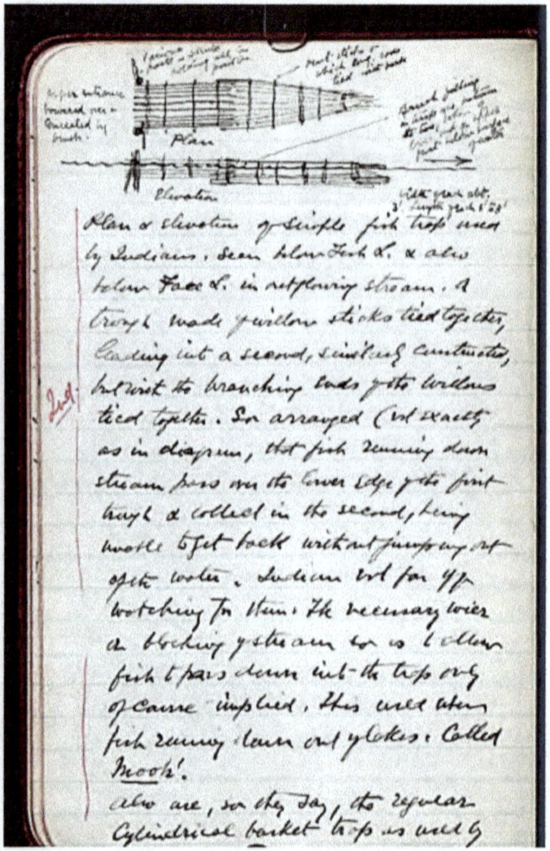

Dawson 1891 McGill niversity
MG1022 6 63 34

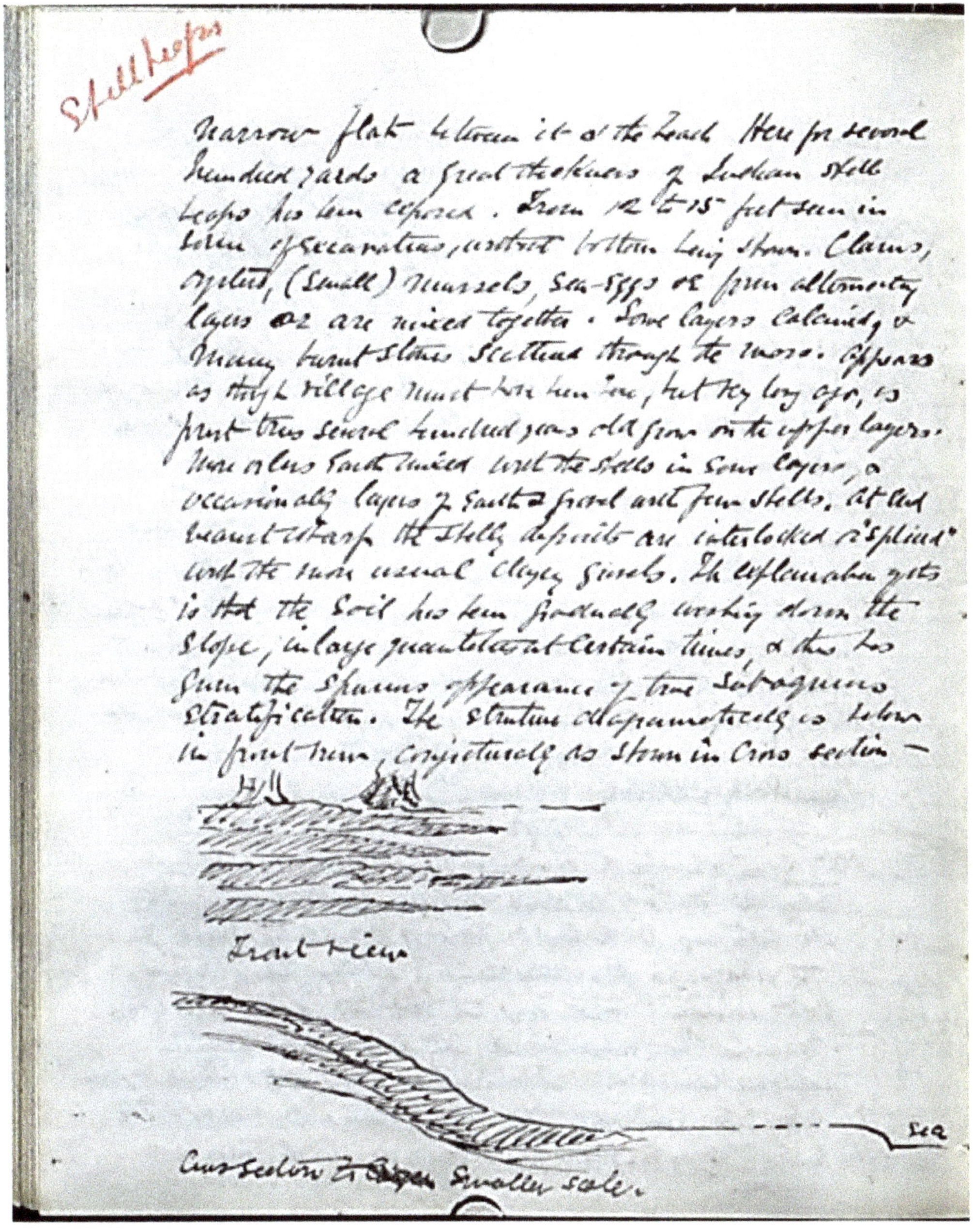

Shell heaps

Narrow flat between it & the beach. Here for several hundred yards a great thickness of Indian shell heaps has been exposed. From 12 to 15 feet seen in face of excavation, without bottom being shown. Clams, oysters, (small) mussels, sea-eggs &c form alternating layers or are mixed together. Some layers calcined, & many burnt stones scattered through the mass. Appears as though village must have been here, but very long ago, as great trees several hundred years old grow on the upper layers. More or less earth mixed with the shells in some layers, & occasionally layers of earth & gravel with few shells. At end nearest Wharf the shelly deposits are interlocked & "splined" with the more usual clayey gravels. The explanation of this is that the soil has been gradually working down the slope, in large quantities at certain times, & thus has given the spurious appearance of true subaqueous stratification. The structure diagrammatically is below in front view, conjecturally as shown in cross section —

Dawson McGill niversity MG1022 6 218

Geology

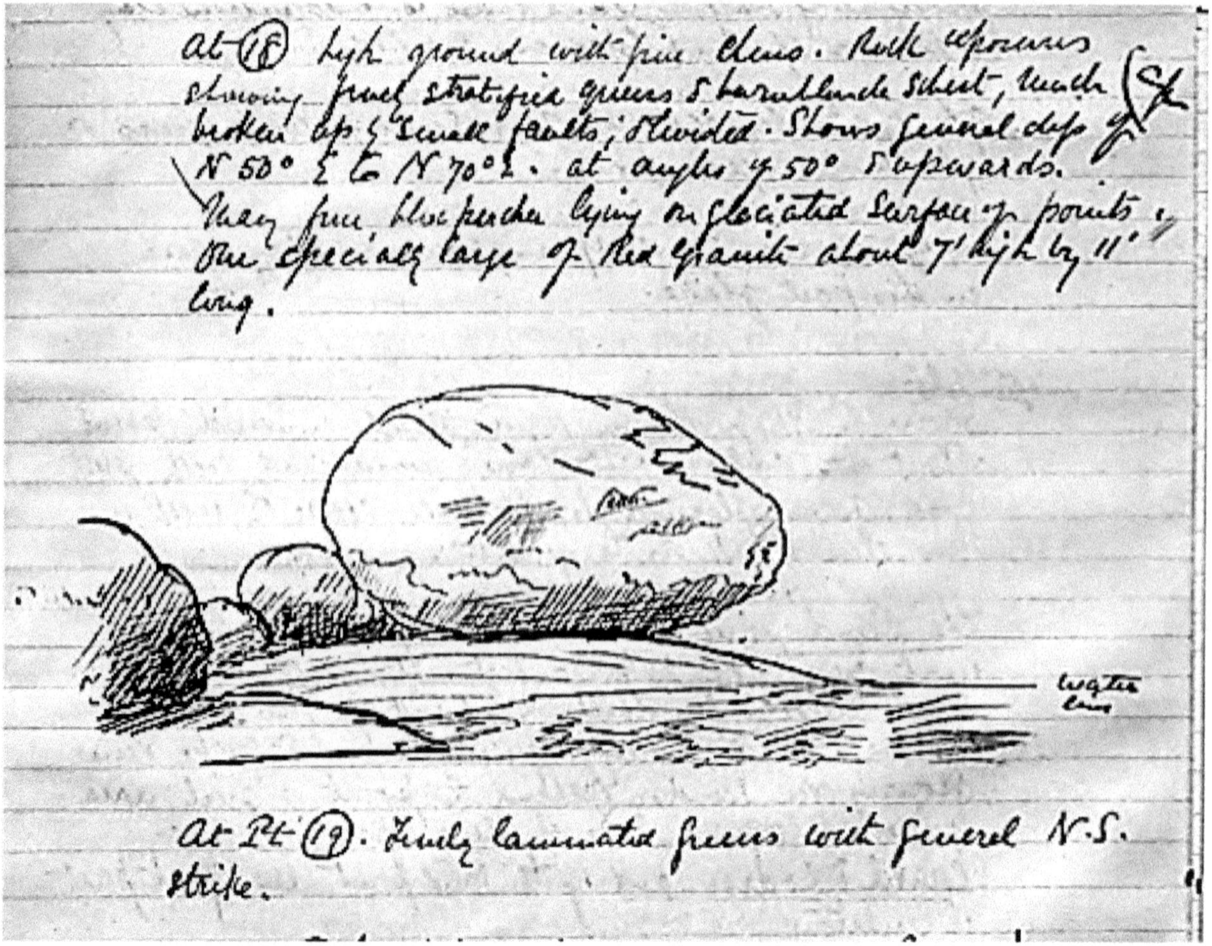
at (18) high ground with pine clumps. Rock exposures showing finely stratified gneiss & hornblende schist, much broken up by small faults, & twisted. Shows general dip of N 50° E to N 70° E. at angles of 50° & upwards. Many fine blocs perches lying on glaciated surface of points, one specially large of Red Granite about 7' high by 11' long.

At Pt (19). Finely laminated gneiss with general N.S. strike.

Dawson 1873. McGill niversity 1022 6 178 274

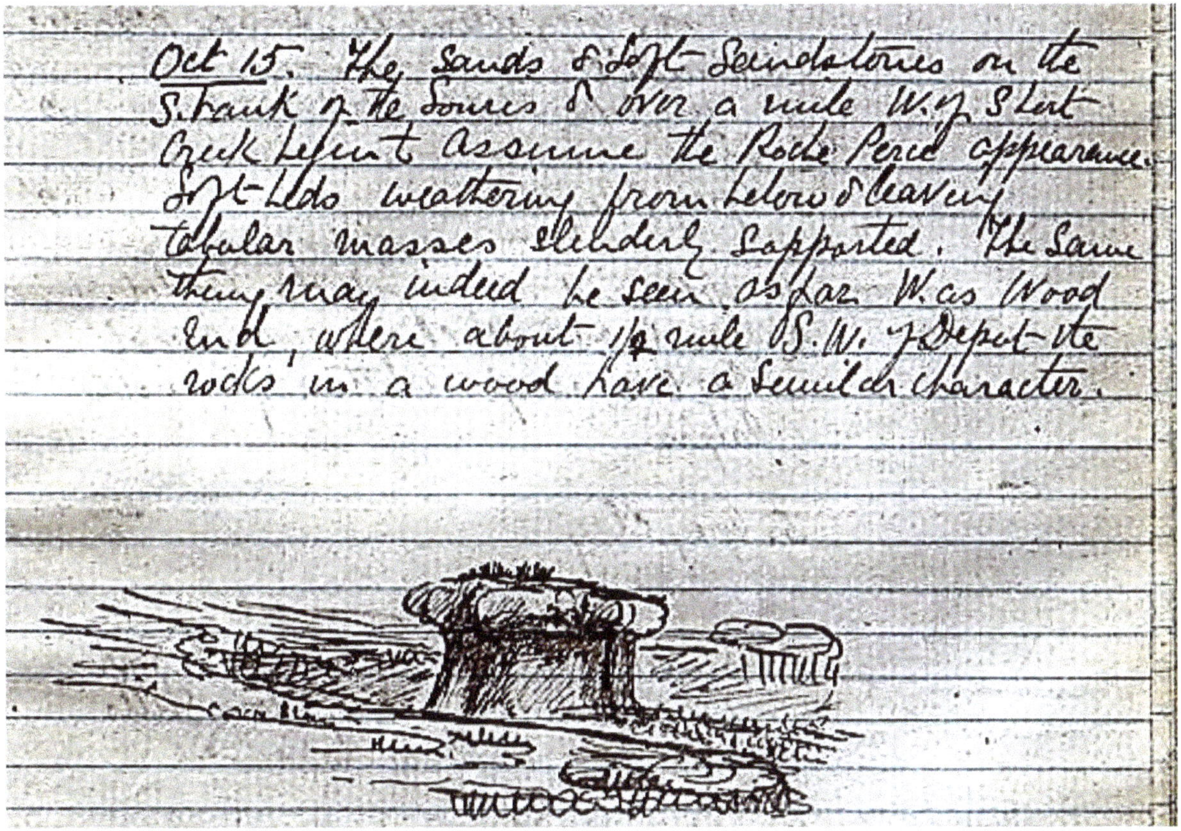

Oct 15. The sands & soft sandstones on the S. bank of the Souris & over a mile W. of Short Creek Depot assume the Roche Percé appearance. Soft beds weathering from below & leaving tabular masses slenderly supported. The same thing may indeed be seen as far W. as Wood End, where about 1/2 mile S.W. of Depot the rocks in a wood have a similar character.

Dawson 1873. McGill niversity 1022 6 177 248

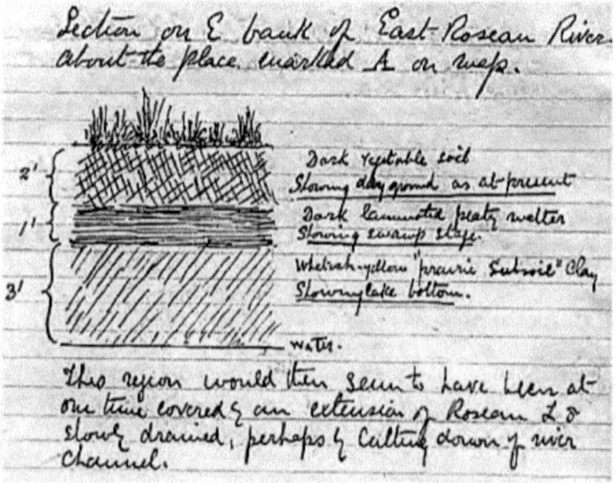

Dawson 1873. McGill niversity
1022 6 178 223

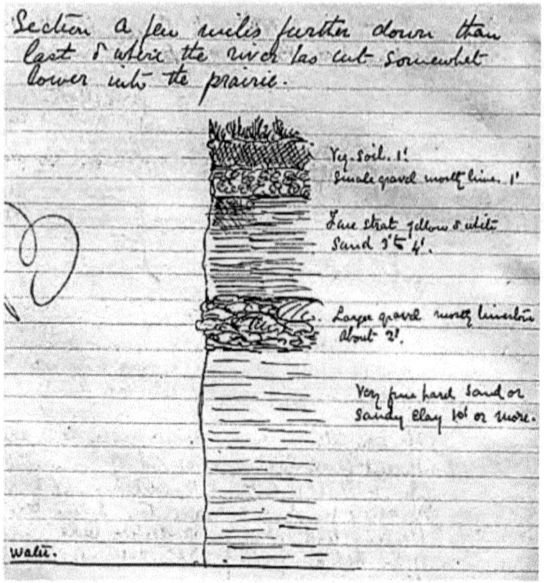

Dawson 1873. McGill niversity
1022 6 178 221

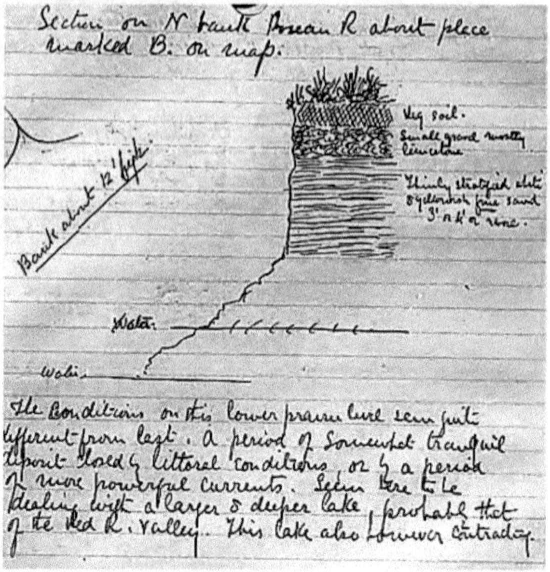

Dawson 1873. McGill niversity
1022 6 178 222

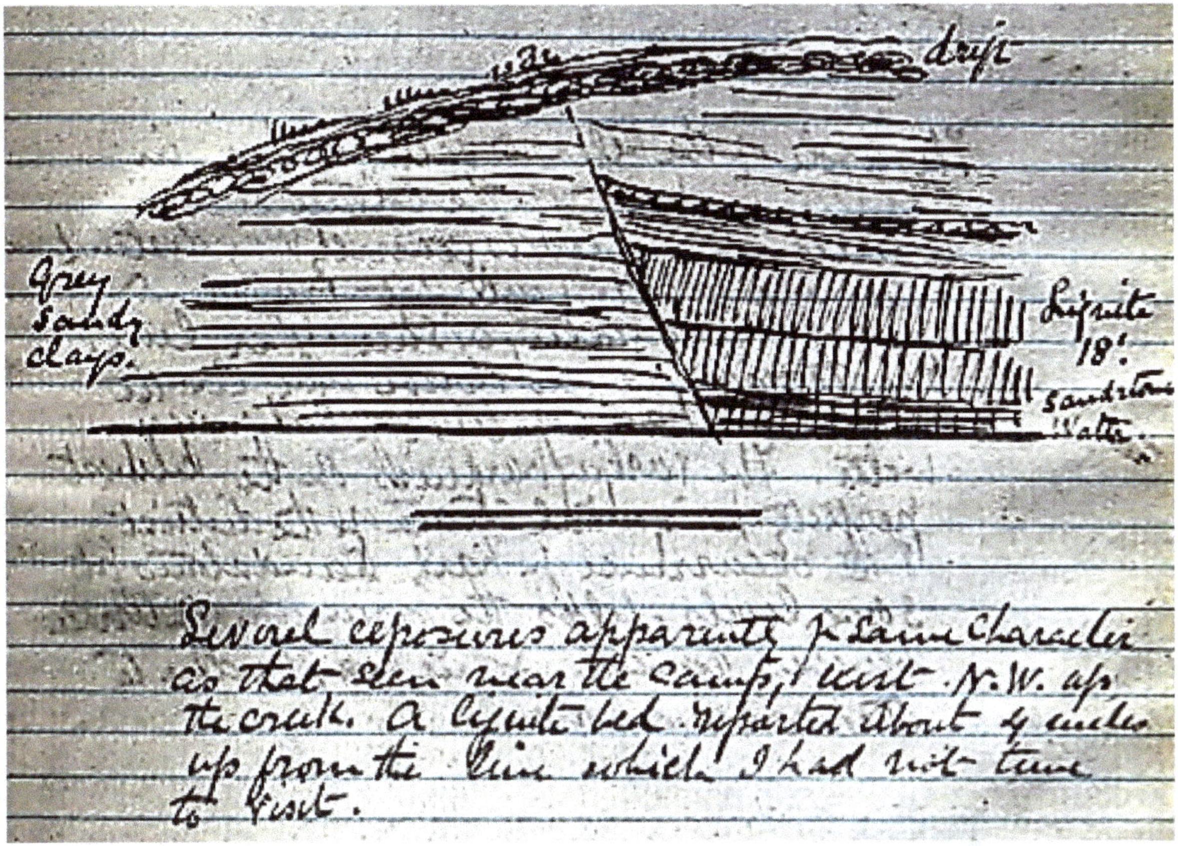

Dawson 1873. McGill niversity 1022 6 177 258

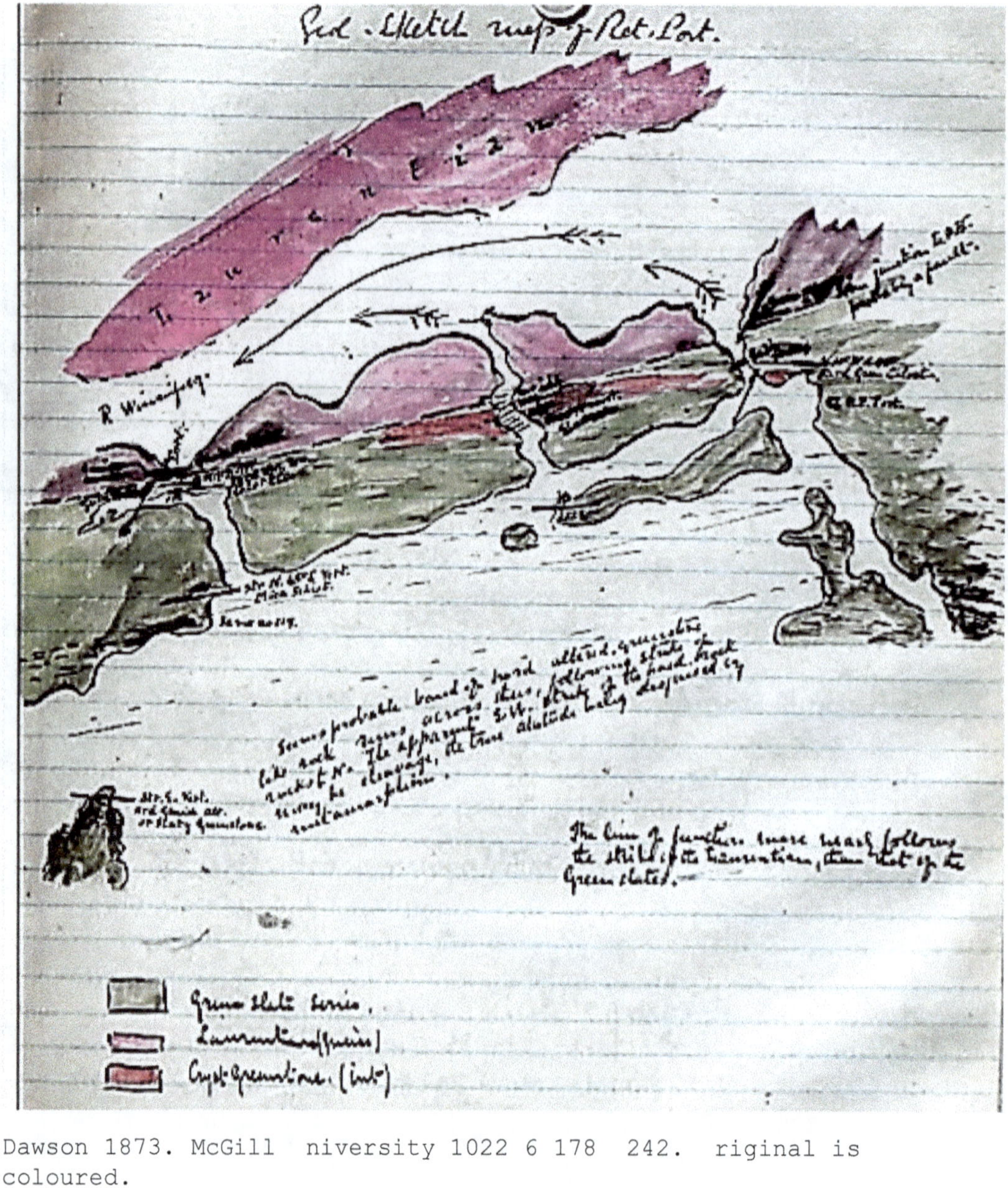

Dawson 1873. McGill niversity 1022 6 178 242. riginal is coloured.

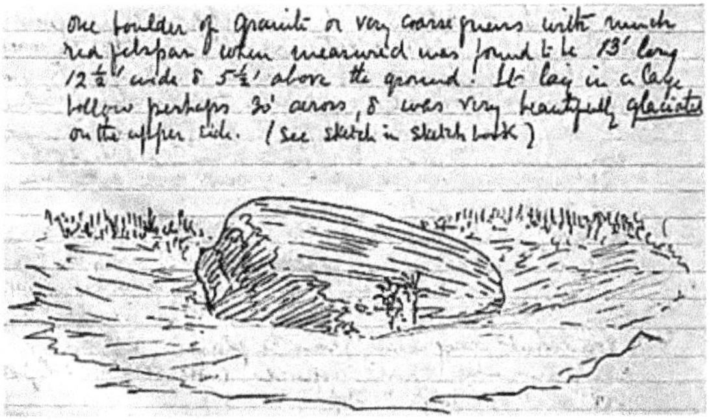

Dawson 1873. McGill niversity
1022 6 178 291

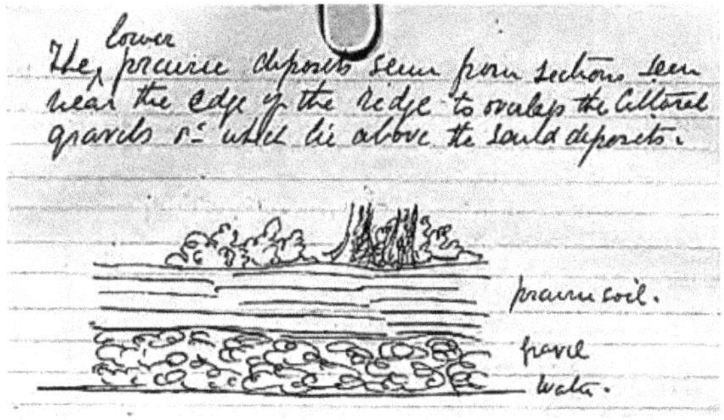

Dawson 1873. McGill niversity
1022 6 178 218 a

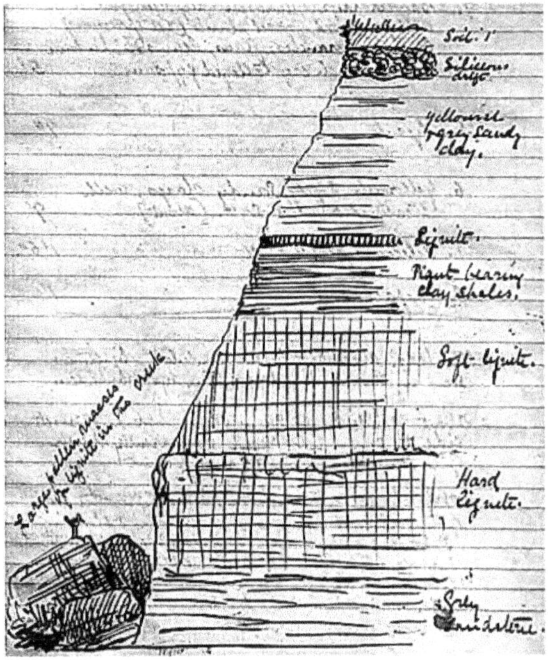

Dawson 1873. McGill niversity
1022 6 177 259

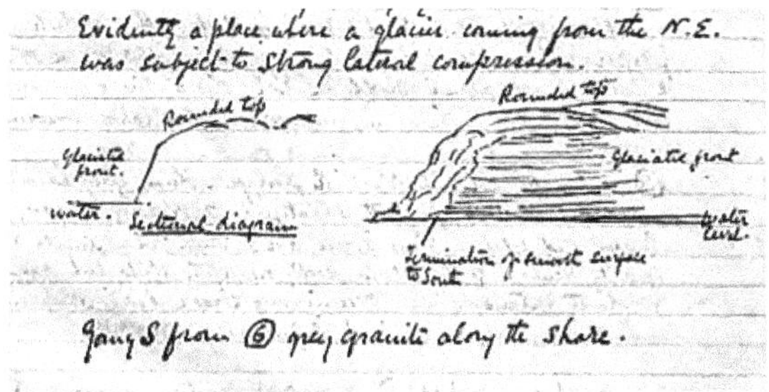

Dawson 1873. McGill niversity
1022 6 178 283

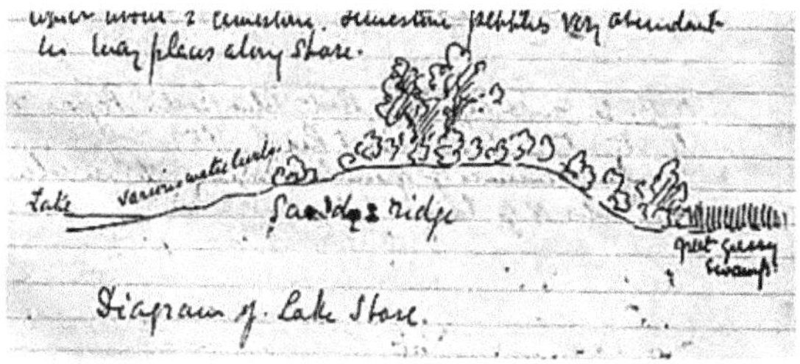

Dawson 1873. McGill niversity
1022 6 178 279

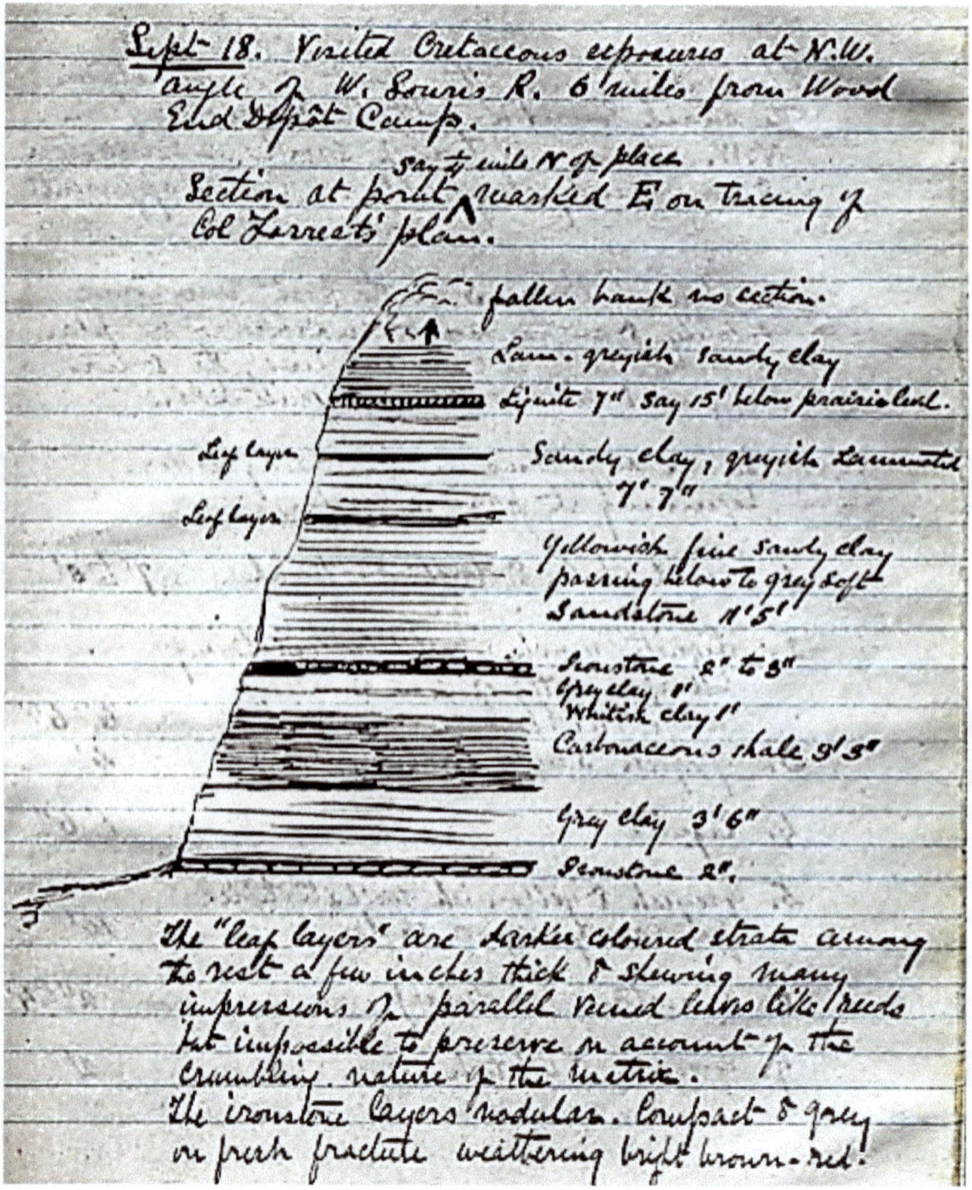

Sept 18. Visited Cretaceous exposures at N.W. angle of W. Souris R. 6 miles from Wood End Depôt Camp.

Section at point say ¼ mile N of place marked E on tracing of Col Forrest's plan.

The "leaf layers" are darker coloured strata among the rest a few inches thick & shewing many impressions of parallel veined leaves like reeds but impossible to preserve on account of the crumbling nature of the matrix.
The ironstone layers nodular. Compact & grey on fresh fracture weathering bright brown-red.

Dawson 1873. McGill niversity 1022 6 177 284

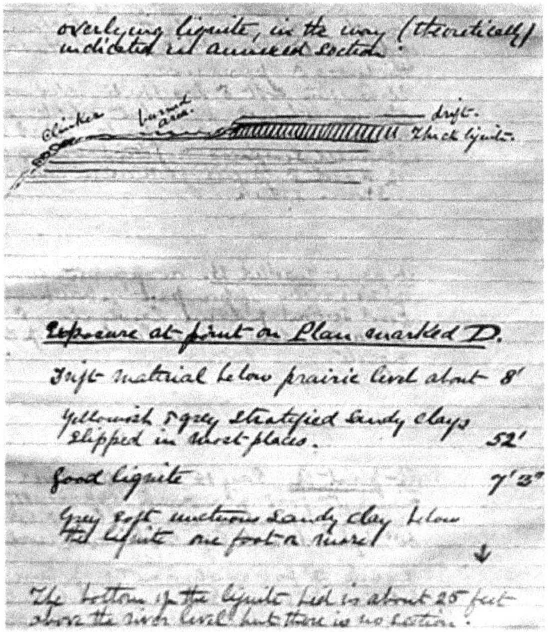

Dawson 1873. McGill niversity
1022 6 177 280

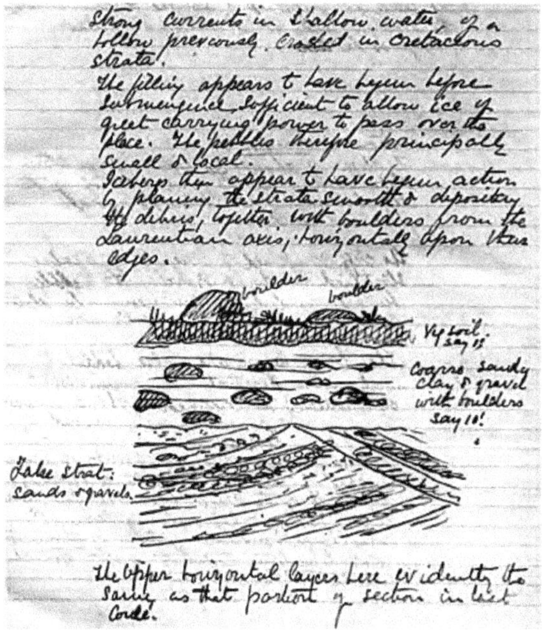

Dawson 1873. McGill niversity
1022 6 177 289

Dawson 1873. McGill niversity
1022 6 177 299

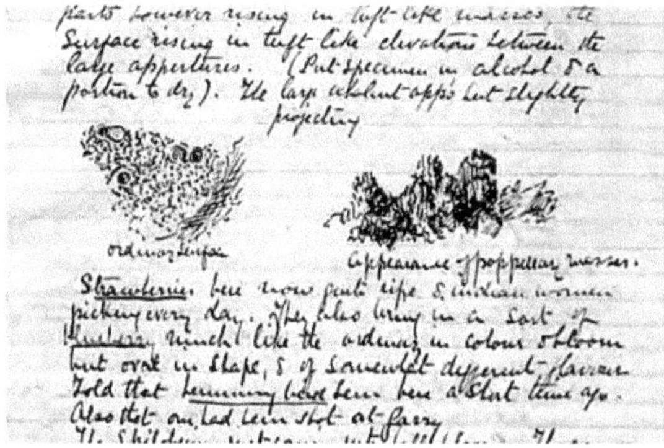

Dawson 1873. McGill niversity
1022 6 178 22

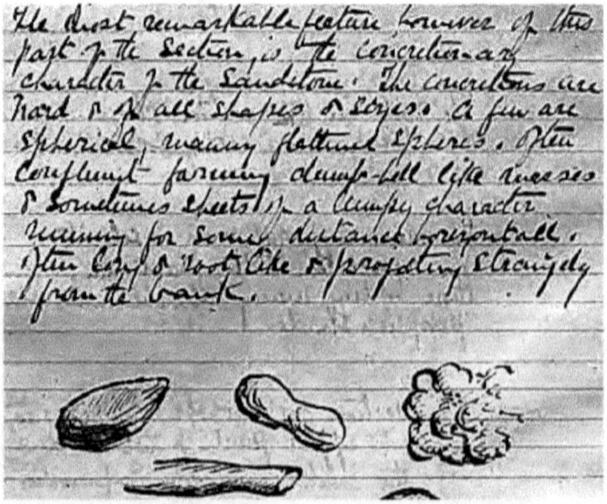
The most remarkable feature however of this part of the Section, is the concretionary character of the Sandstone. The concretions are hard & of all shapes & sizes. A few are spherical, many flattened spheres. Often confluent forming dumb-bell like masses & sometimes sheets of a lumpy character running for some distance horizontally. Often long & root like & projecting strangely from the bank.

Dawson 1873. McGill niversity
1022 6 177 254

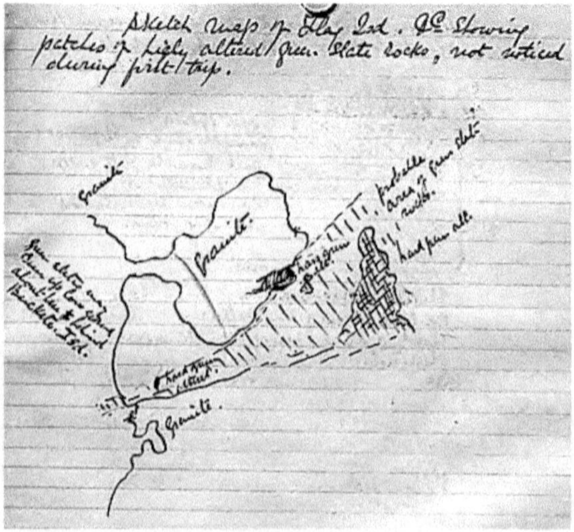
Sketch map of Flag Isd. &c Showing patches of highly altered green slate rocks, not noticed during first trip.

Dawson 1873. McGill niversity
1022 6 178 224

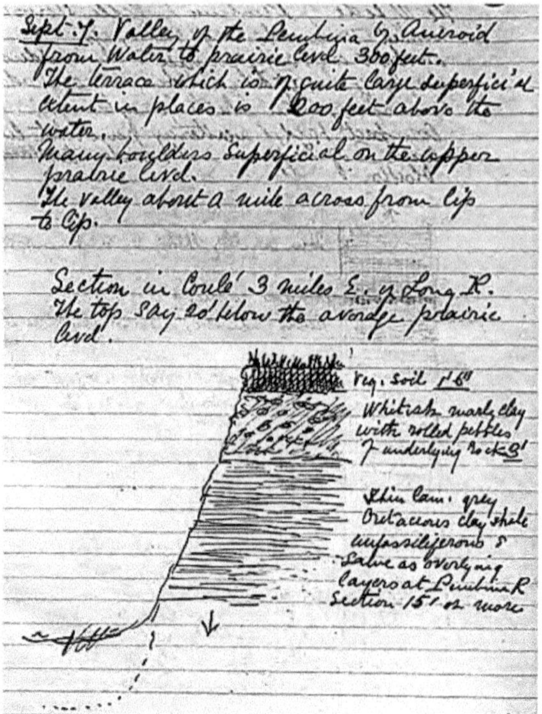
Sept. 7. Valley of the Pembina by Aneroid from Water to prairie level 300 feet.
The terrace which is of quite large superficial extent in places is 200 feet above the water.
Many boulders superficial on the upper prairie level.
The valley about a mile across from lip to lip.

Section in Coulé 3 miles E. of Long R.
The top say 20' below the average prairie level.

Dawson 1873. McGill niversity
1022 6 177 291

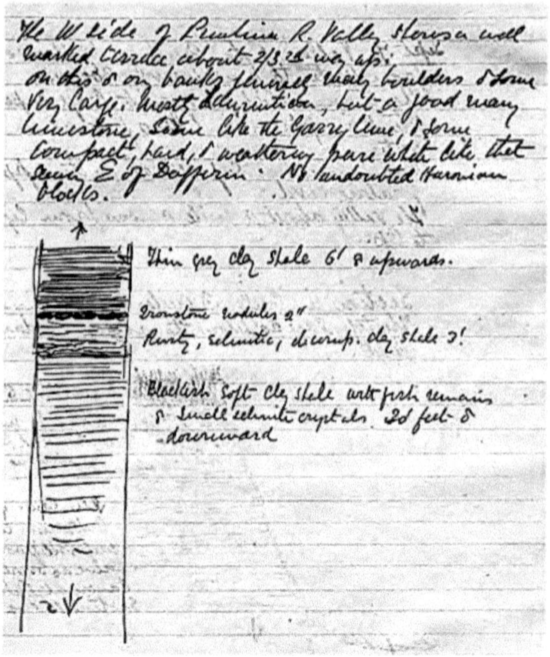
The W. side of Pembina R. Valley shows a well marked terrace about 2/3 rds way up.
On this & on banks generally many boulders & some very large. Mostly Laurentian, but a good many limestone, some like the Garry lime, & some compact, hard, & weathering pure white like that seen E of Dufferin. No undoubted Huronian blocks.

Dawson 1873. McGill niversity
1022 6 177 292

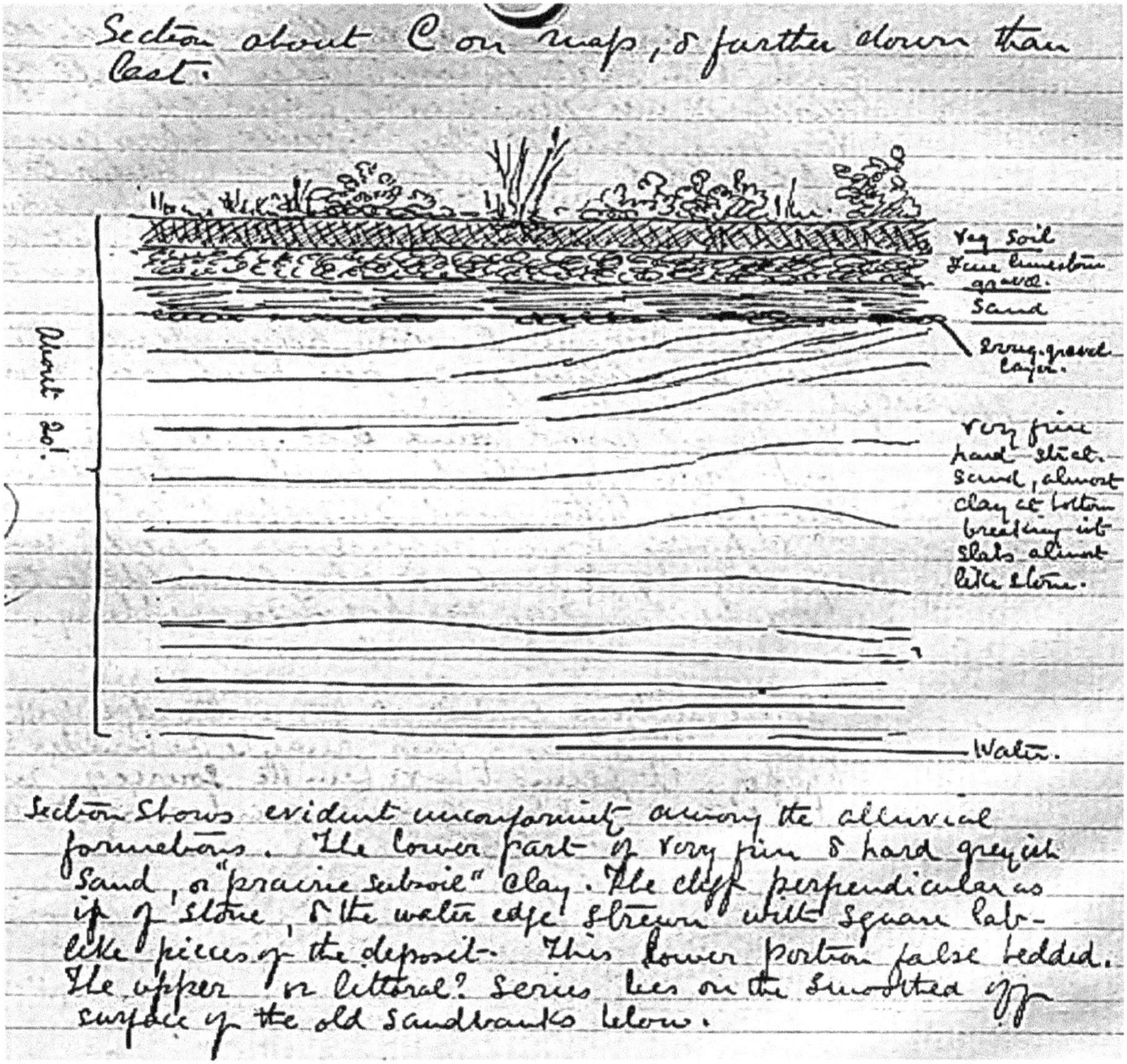

Dawson 1873. McGill niversity 1022 6 178 220

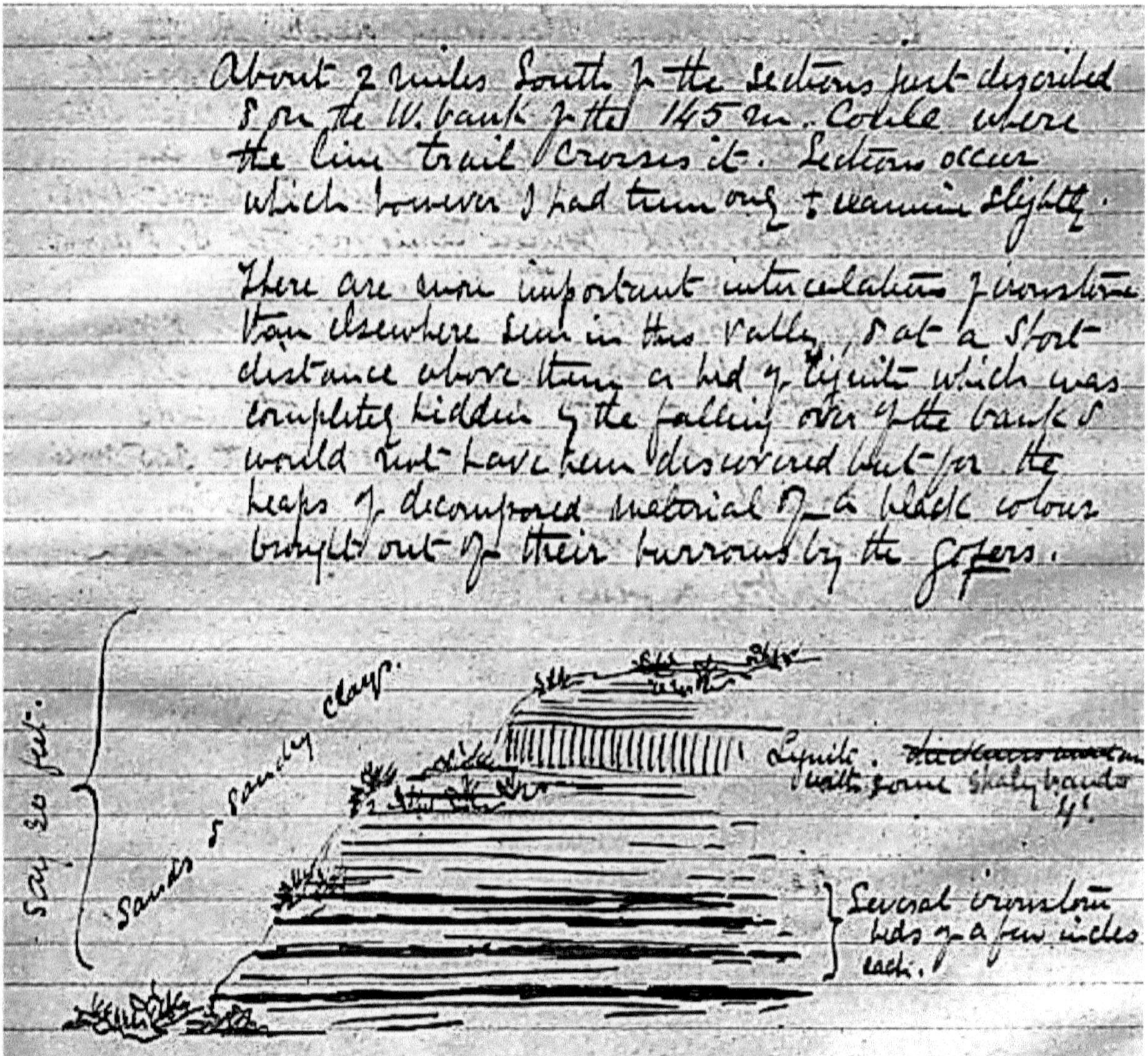
About 2 miles South of the sections just described & on the W. bank of the 145 m. Coulee where the line trail crosses it. Sections occur which however I had time only to examine slightly.

There are more important intercalations of ironstone than elsewhere seen in this valley, & at a short distance above them a bed of lignite which was completely hidden by the falling over of the bank & would not have been discovered but for the heaps of decomposed material of a black colour brought out of their burrows by the gophers.

Dawson 1873. McGill niversity 1022 6 177 266

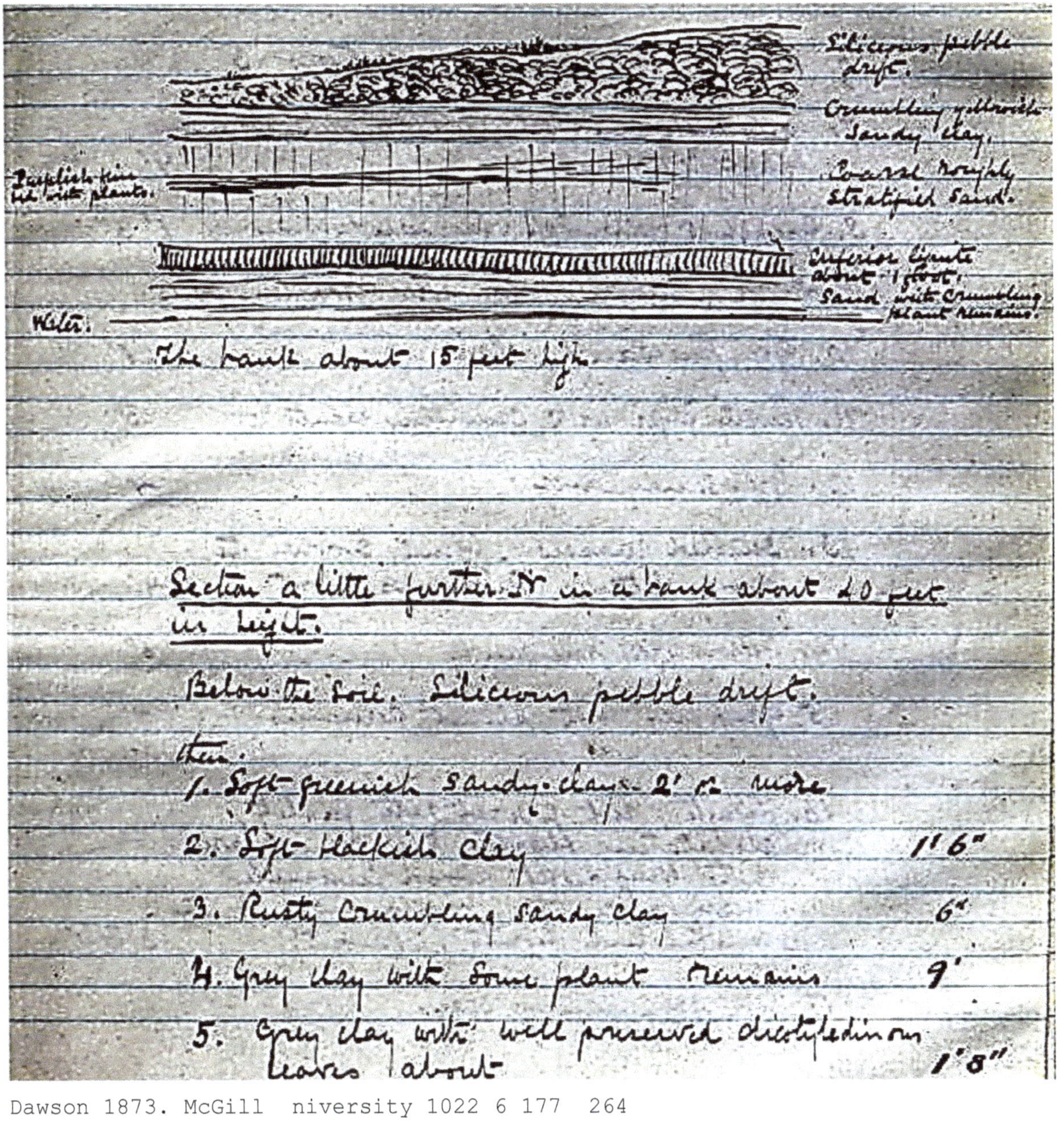

The bank about 15 feet high.

Section a little farther N in a bank about 20 feet in height.

Below the soil. Siliceous pebble drift.

then.

1. Soft greenish sandy clay 2' or more
2. Soft blackish clay 1'6"
3. Rusty crumbling sandy clay 6"
4. Grey clay with some plant remains 9'
5. Grey clay with well preserved dicotyledinous leaves about 1'8"

Dawson 1873. McGill niversity 1022 6 177 264

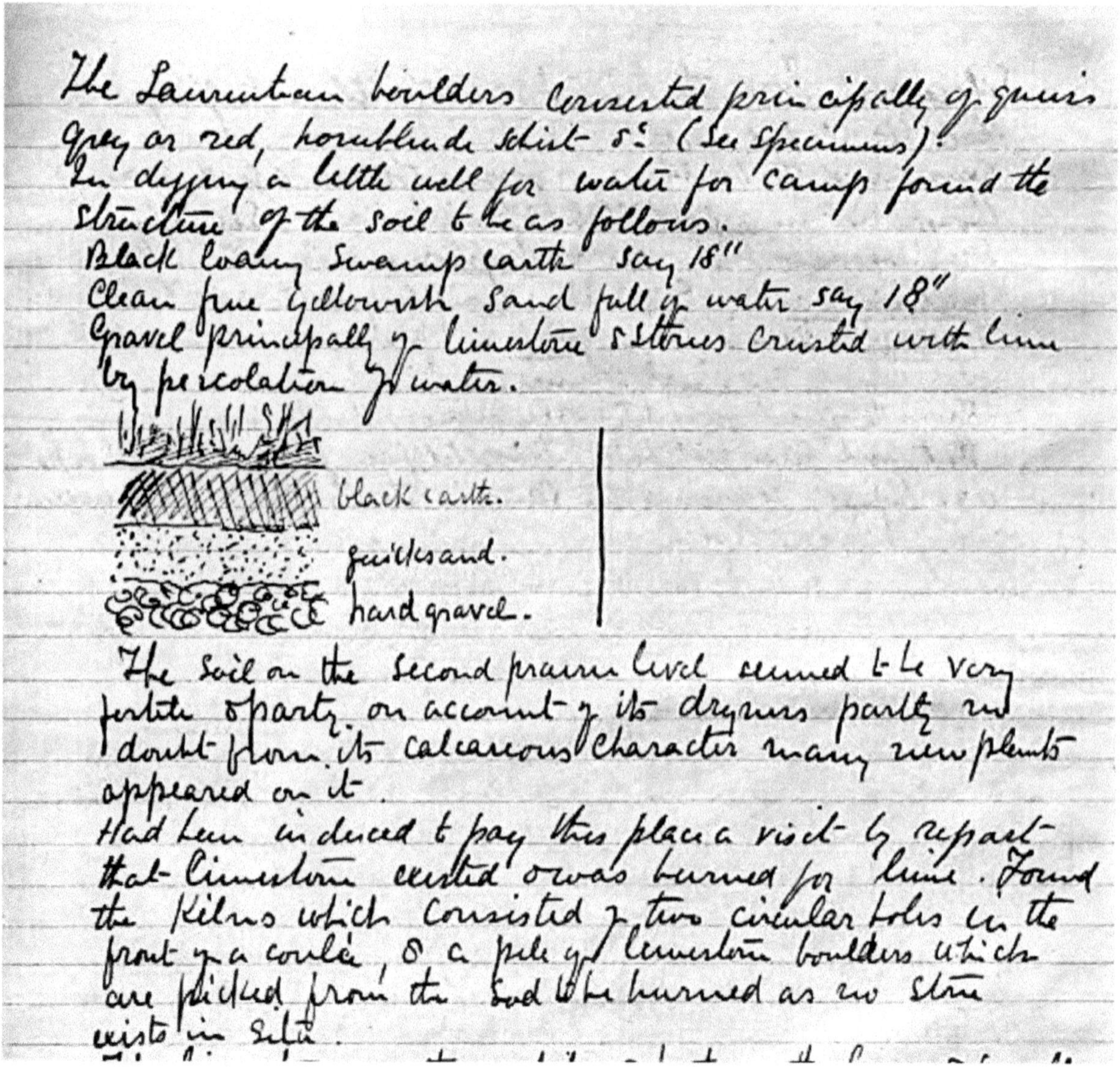

The Laurentian boulders consisted principally of gneiss grey or red, hornblende schist &c (See Specimens).
In digging a little well for water for camp found the structure of the soil to be as follows.
Black loamy swamp earth say 18"
Clean fine yellowish sand full of water say 18"
Gravel principally of limestone & stones crusted with lime by percolation of water.

The soil on the second prairie level seemed to be very fertile & partly on account of its dryness partly no doubt from its calcareous character many new plants appeared on it.
Had been induced to pay this place a visit by report that limestone existed & was burned for lime. Found the kilns which consisted of two circular holes in the front of a coulée, & a pile of limestone boulders which are picked from the sod to be burned as no stone exists in situ.

Dawson 1873. McGill niversity 1022 6 178 290

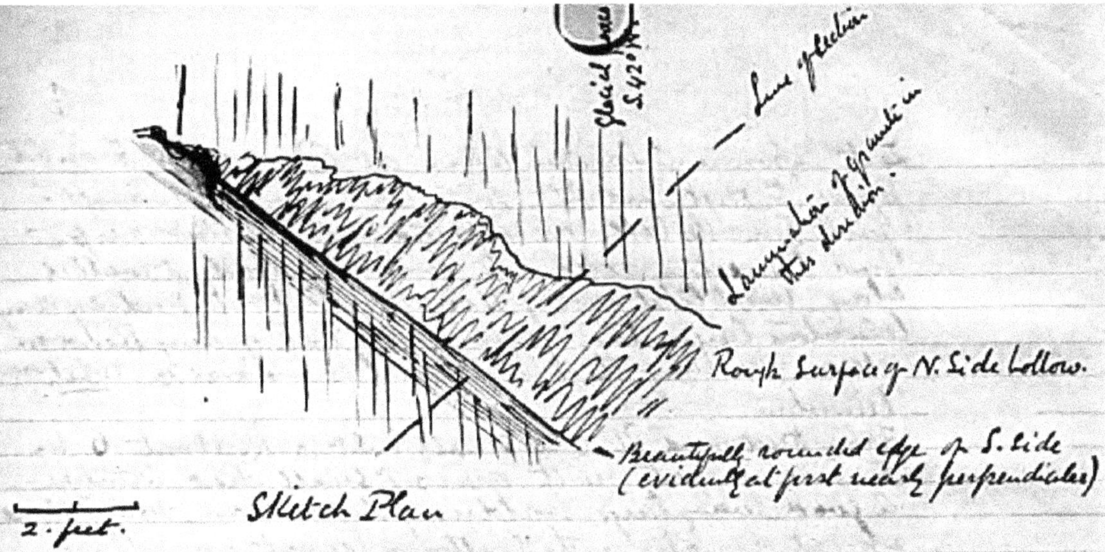

Sketch Plan

Section along dotted line.

Got specimens of coarse greenstone with pyrites specks from two islands lying off here from ½ m to 1 m S about place marked (B) on map.

N.B. See Hinds Report in Brit. Blue Book. "Papers relating to country between L. Superior & Red R. 1859.
Mentions occurrence of "greenstone conglomerate" on island 20 or 25? miles N of Garden Island.

Dawson 1873. McGill niversity 1022 6 178 280

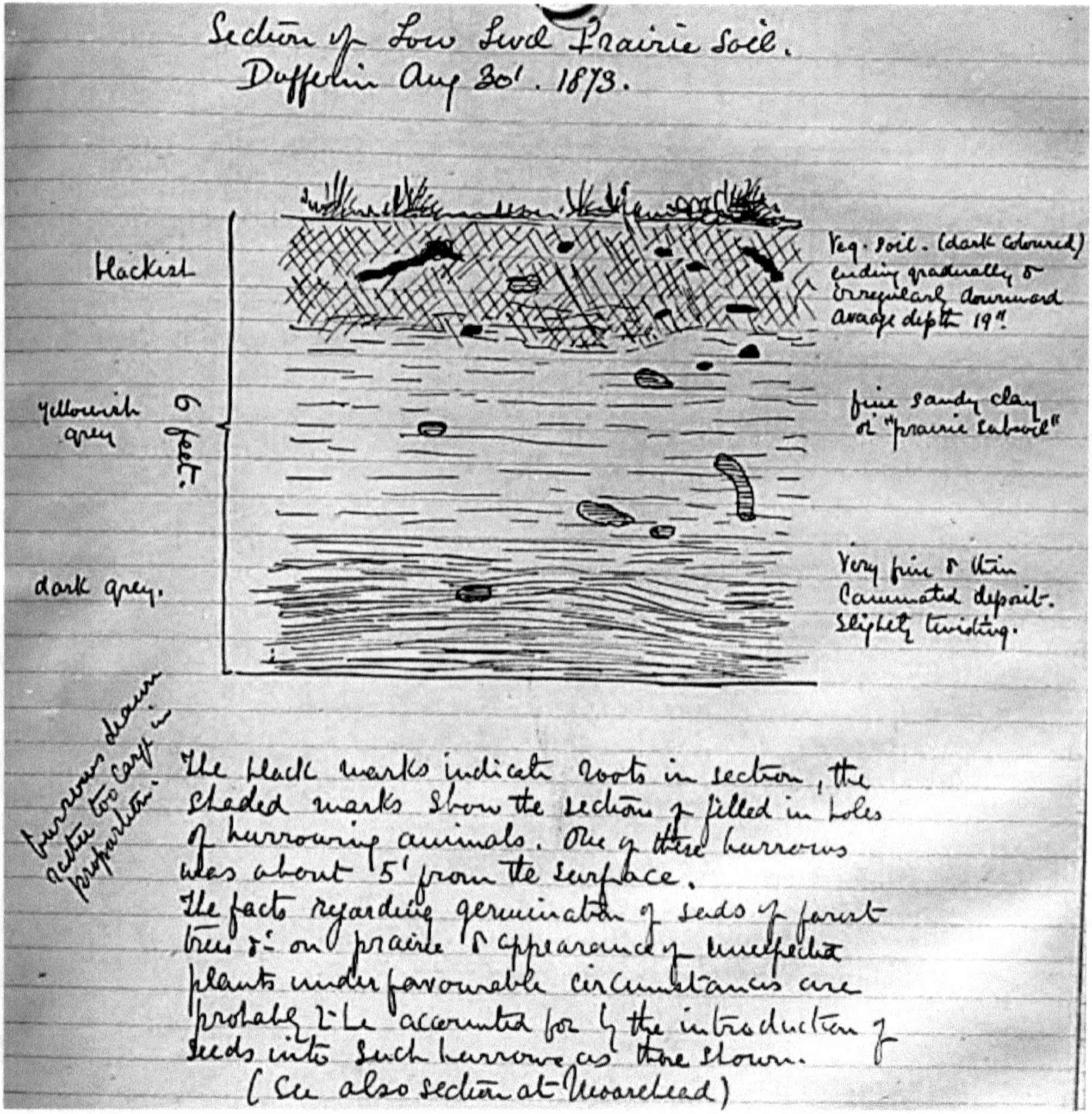

Dawson 1873. McGill niversity 1022 6 178 216

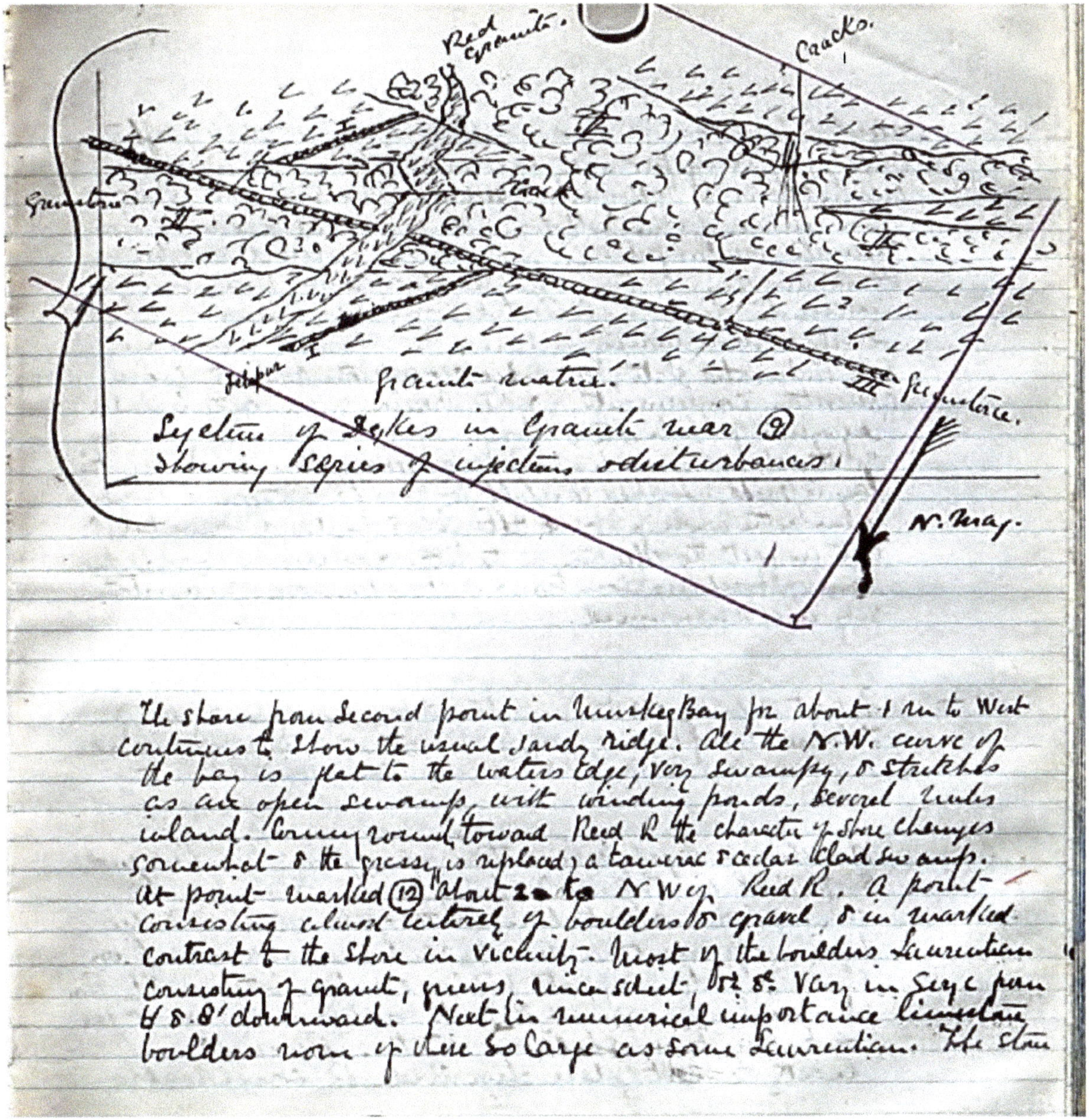

The shore from Second point in Muskeg Bay for about 1 m to West continues to show the usual sandy ridge. All the N.W. curve of the bay is flat to the waters edge, very swampy, & stretches as an open swamp, with winding ponds, several miles inland. Coming round toward Reed R the character of shore changes somewhat & the grassy is replaced a tamarac & cedar clad swamp. At point marked (12) about 2 m to N.W. of Reed R. a point consisting almost entirely of boulders & gravel & in marked contrast to the shore in vicinity. Most of the boulders Laurentian consisting of granite, gneiss, mica schist, &c &c. Vary in size from 4 & 5' downward. Next in numerical importance limestone boulders none of these so large as some Laurentian. The stone

Dawson 1873. McGill niversity 1022 6 178 278

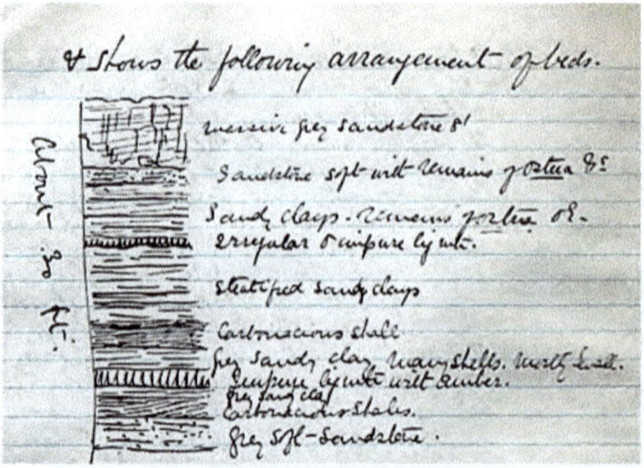

Dawson 1874. McGill niversity 1022 6
180 128

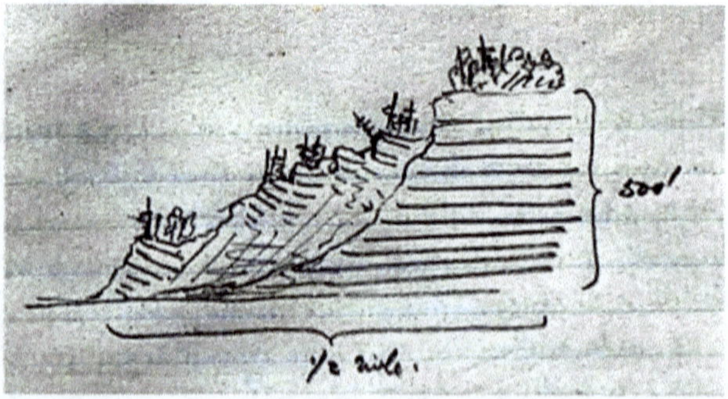

Dawson 1879. McGill niversity 1022 6
135 174

Dawson 1874. McGill niversity 1022 6
180 174

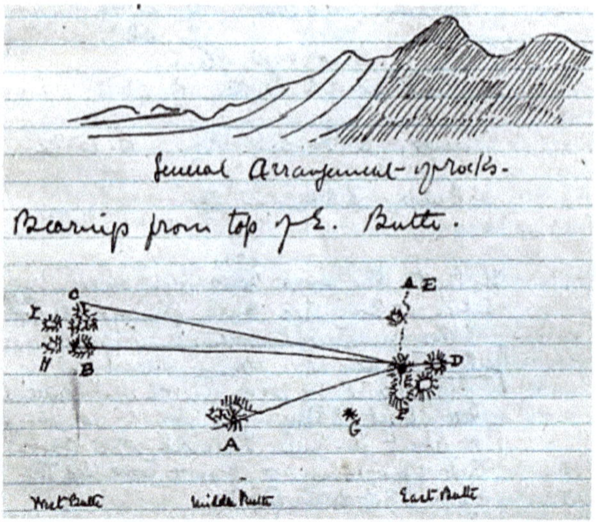

Dawson 1874. McGill niversity 1022 6
180 133

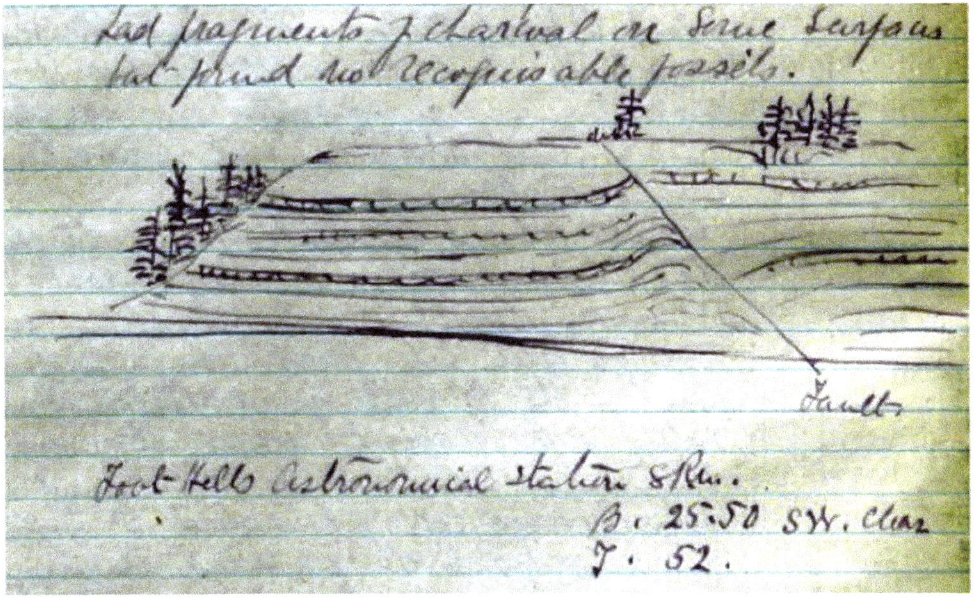

Dawson 1874. McGill niversity 1022 6 180 166

Dawson 1874. McGill niversity 1022 6 180 100

Dawson 1875. Geological Survey of Canada field notebooks.

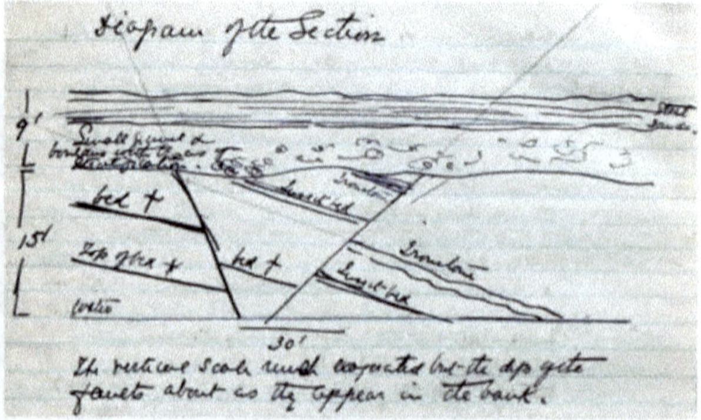

Dawson 1875. McGill niversity 1022 6 182 19

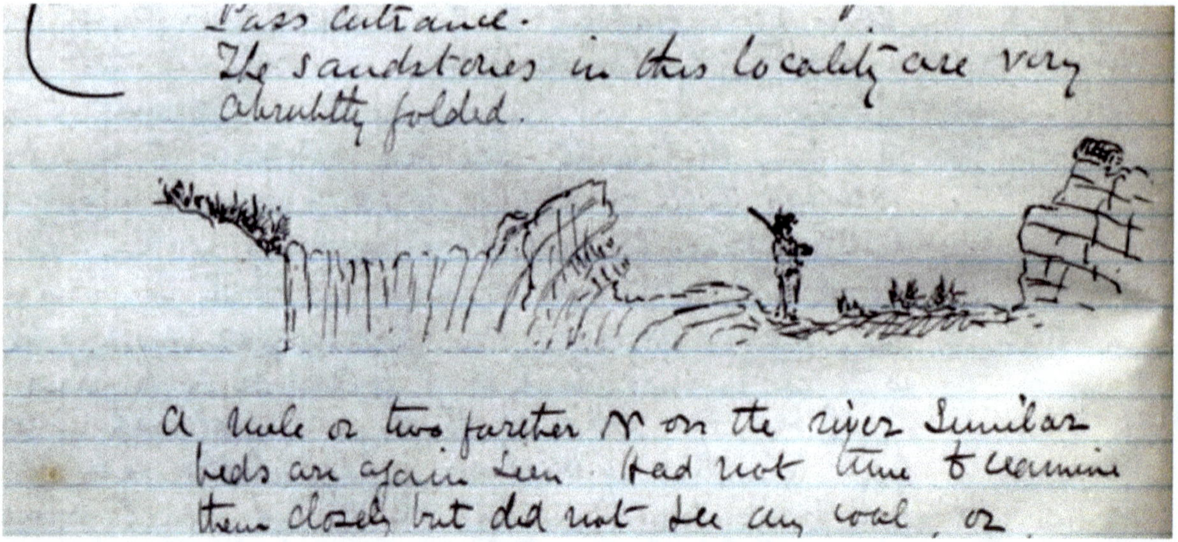

Dawson 1874. McGill niversity 1022 6 180 216

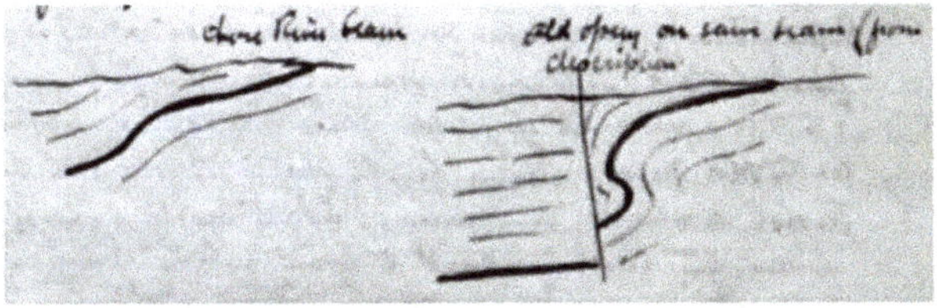

Dawson. 1879. McGill niversity 1022
6   220

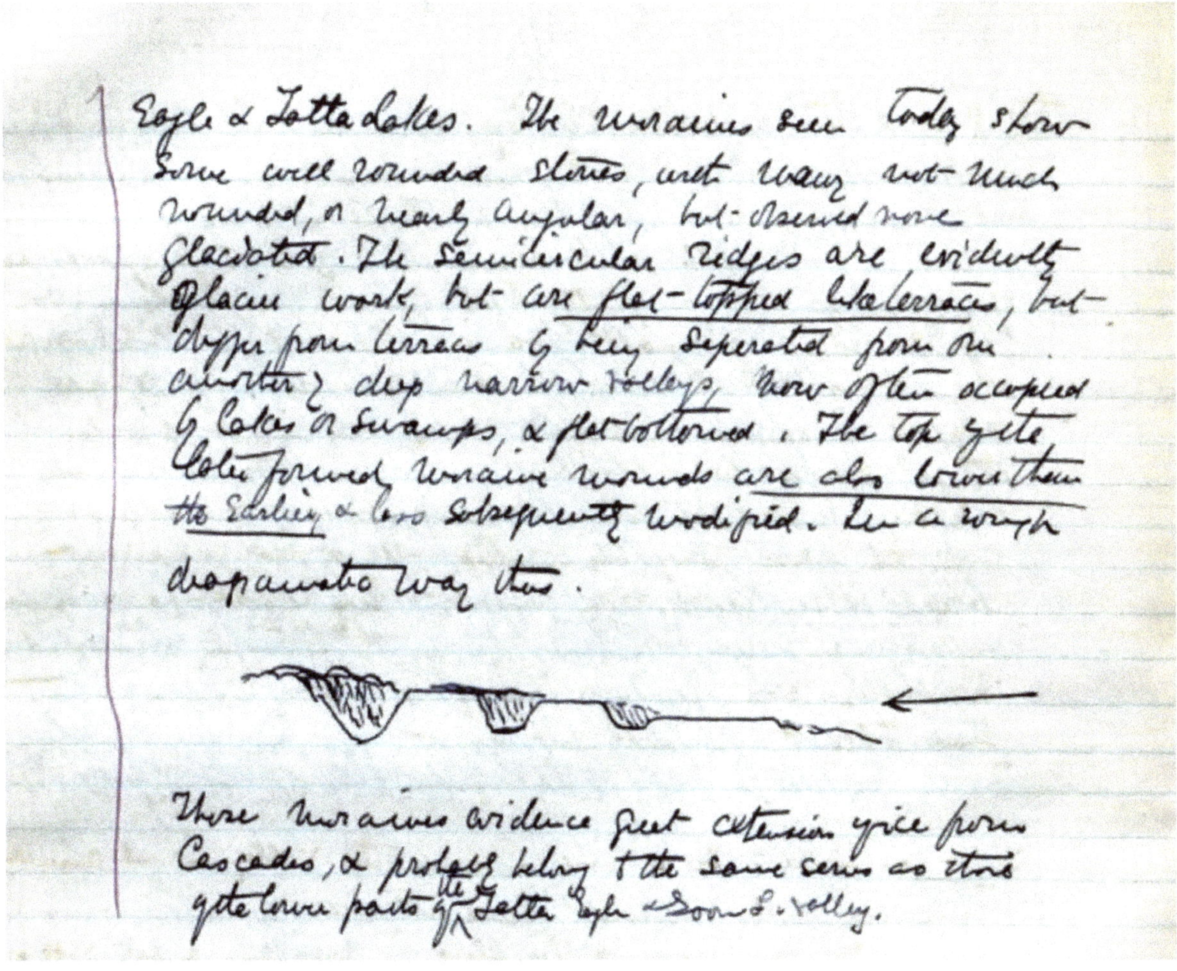
Eagle & Totta Lakes. The moraines seen today show some well rounded stones, with many not much rounded, or nearly angular, but observed none glaciated. The semicircular ridges are evidently glacier work, but are flat-topped like terraces, but differ from terraces by being separated from one another by deep narrow valleys, now often occupied by lakes or swamps, & flat bottomed. The tops of the later formed moraine mounds are also lower than the earlier & less subsequently modified. See a rough diagrammatic way thus.

These moraines evidence great extension of ice from Cascades, & probably belong to the same series as those of the lower parts of the Totta, Eagle & Doon L. valley.

Dawson 1875. McGill niversity 1022 6 182 36

the sides as there is an edge of strata unchanged intervening between the drift & the baked rocks. The combustion has probably been stayed by the wet state of the rocks near the surface

Roughly thus:

The strata now present all the varieties of altered rock described in my report of the Lignite formation on Boundary Line. Large ironstone balls, in some cases more than three feet in diameter occur in one part of the section, their centres appear to have been radially crystalline carbonate of iron, but are now changed into oxide by the heat, & where exposed are breaking up into long splinters which give out a ringing sound when struck, & look when lying about like heaps of nails.

Dawson 1875. McGill niversity 1022 6 182 14

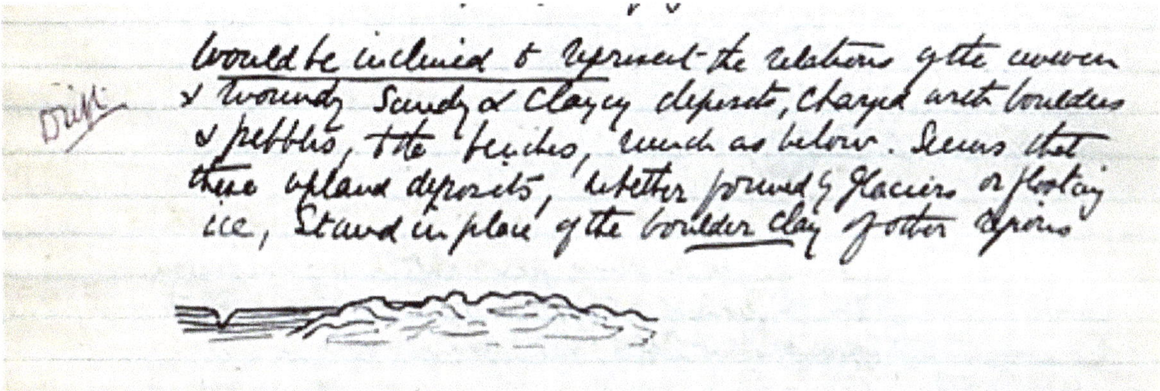

Drift

would be inclined to represent the relations of the uneven & rounded sandy & clayey deposits, charged with boulders & pebbles, to the benches, much as below. Seems that these upland deposits, whether formed by glaciers or floating ice, stand in place of the boulder clay of other regions

Dawson 1875. McGill niversity 1022 6 182   35

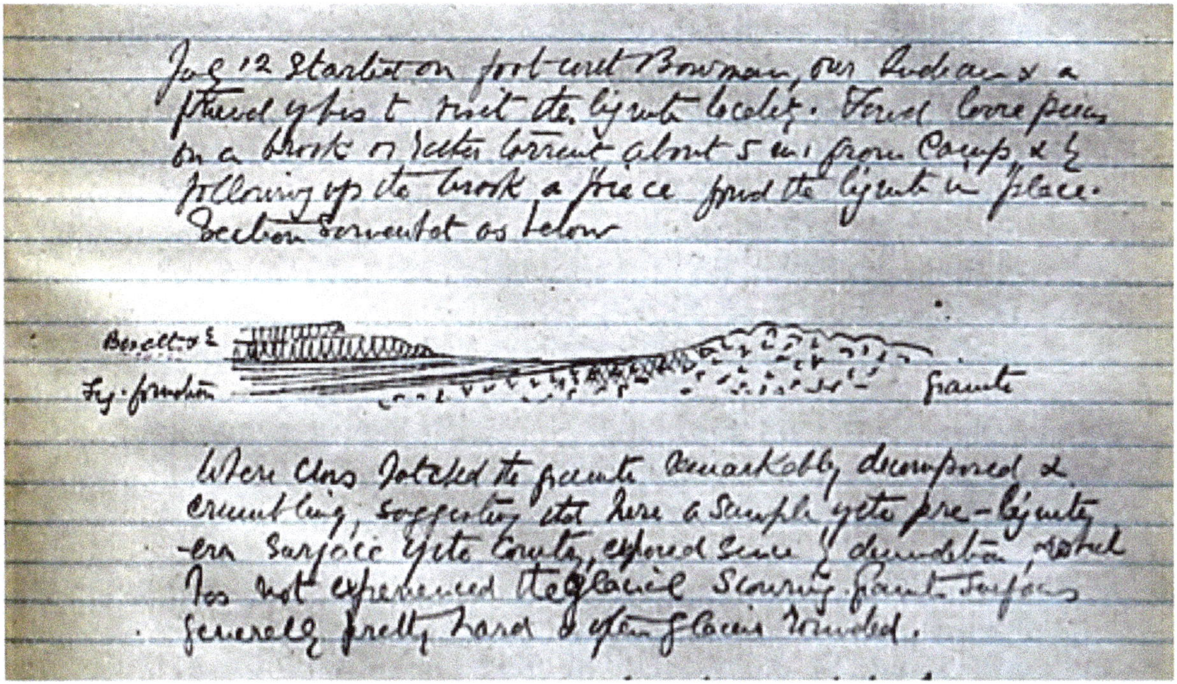

Aug 12 Started on foot with Bowman, our Indians & a friend of his to visit the lignite locality. Found loose pieces in a brook or rather torrent about 5 m. from Camp, & by following up the brook a piece found the lignite in place. Section somewhat as below

Where clays touched the granite remarkably decomposed & crumbling, suggesting that here a sample of the pre-lignite era surface of the country, exposed since by denudation, which has not experienced the glacial scouring. Granite surfaces generally pretty hard & often glacier rounded.

Dawson 1876. McGill niversity 1022 6 131   83

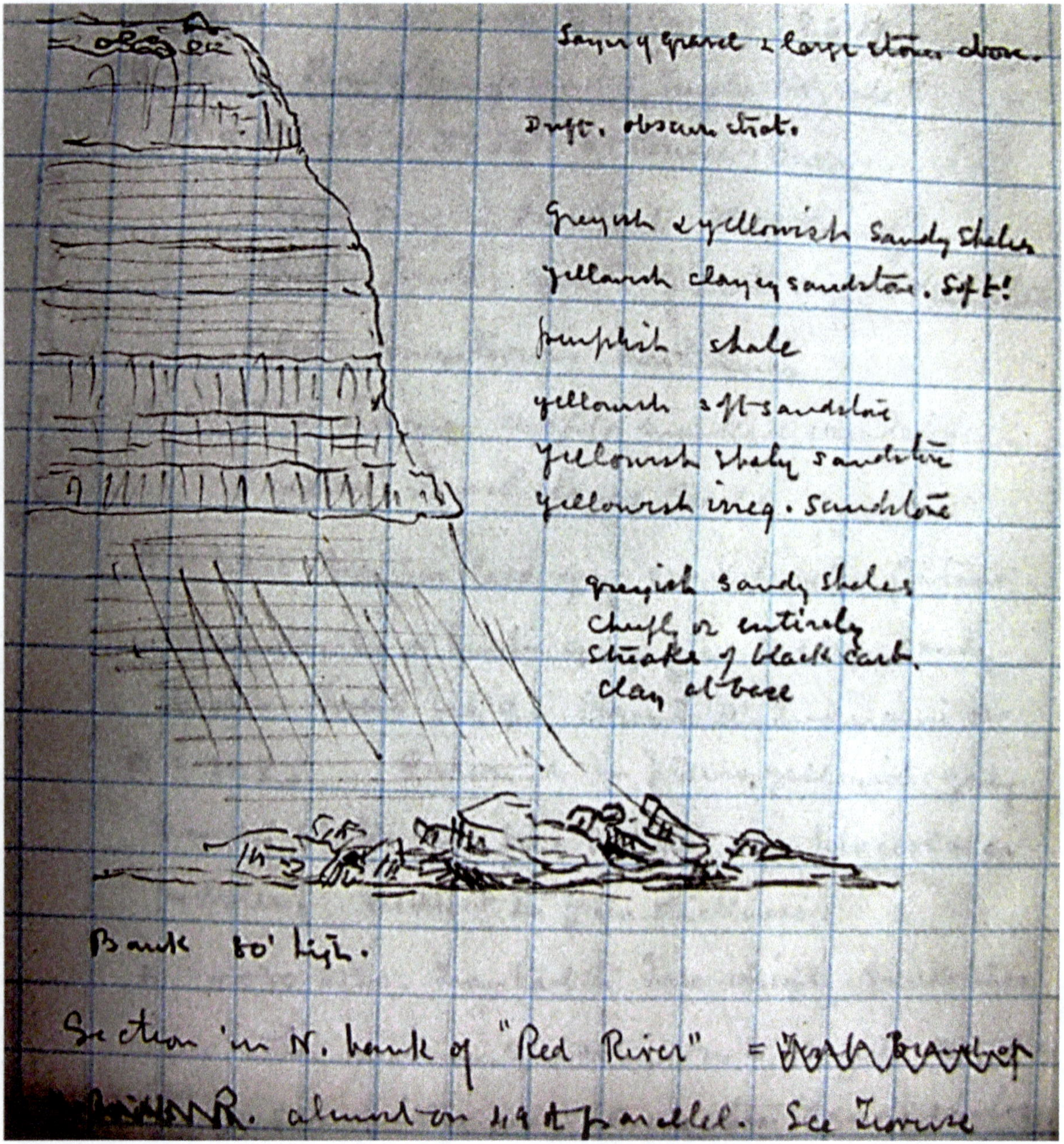

dawson 1879 Geological Survey of Canada field notebooks.

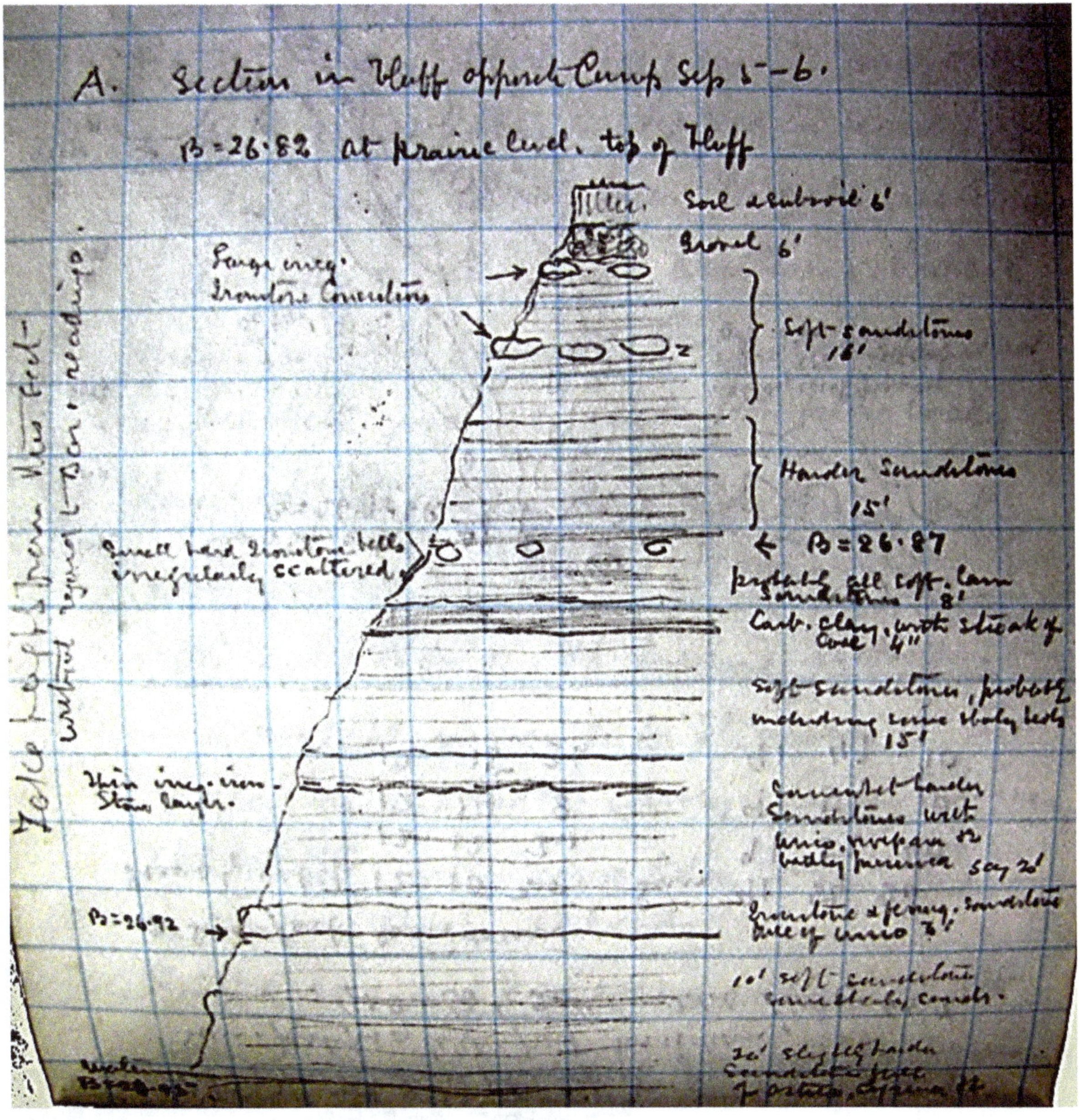

Dawson 1881 no 3 Geological Survey of Canada field notebooks.

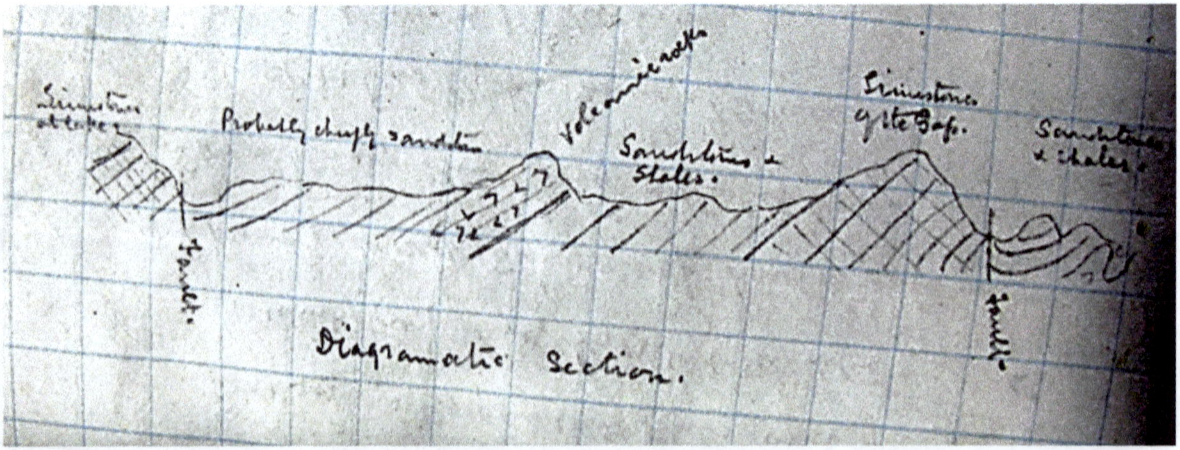

Dawson 1881 no 4 Geological Survey of Canada field notebooks.

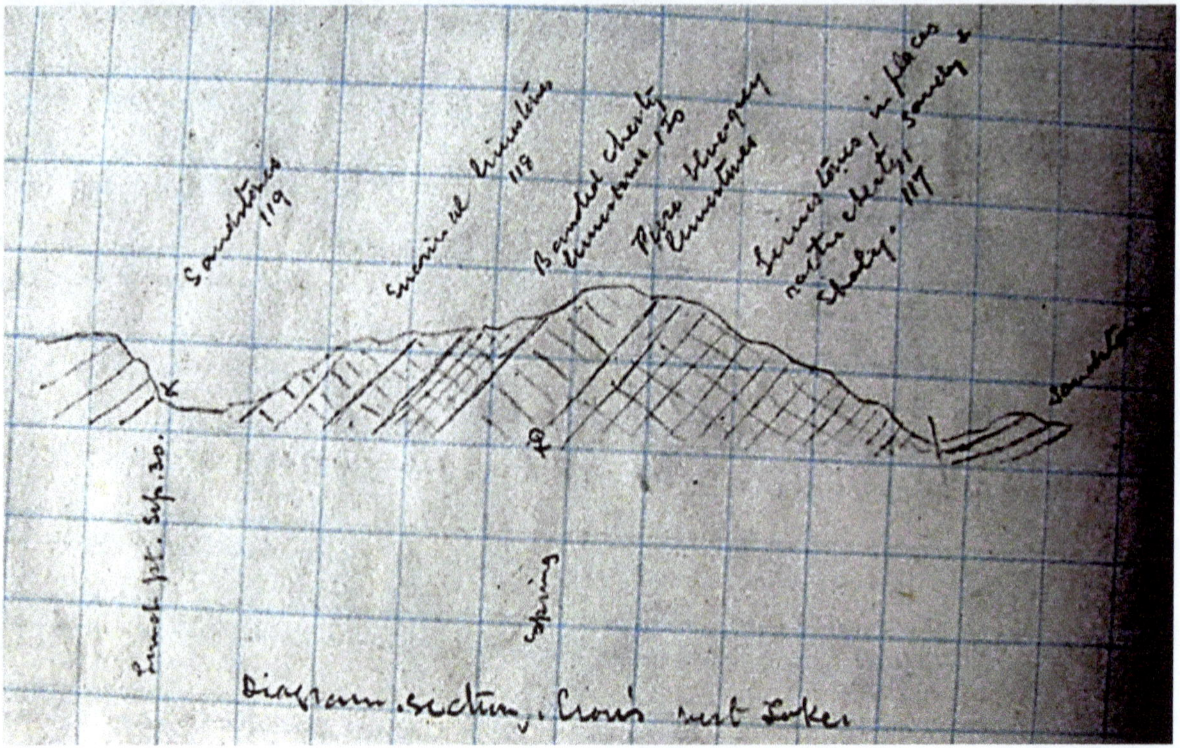

Dawson 1881 no 4 Geological Survey of Canada field notebooks.

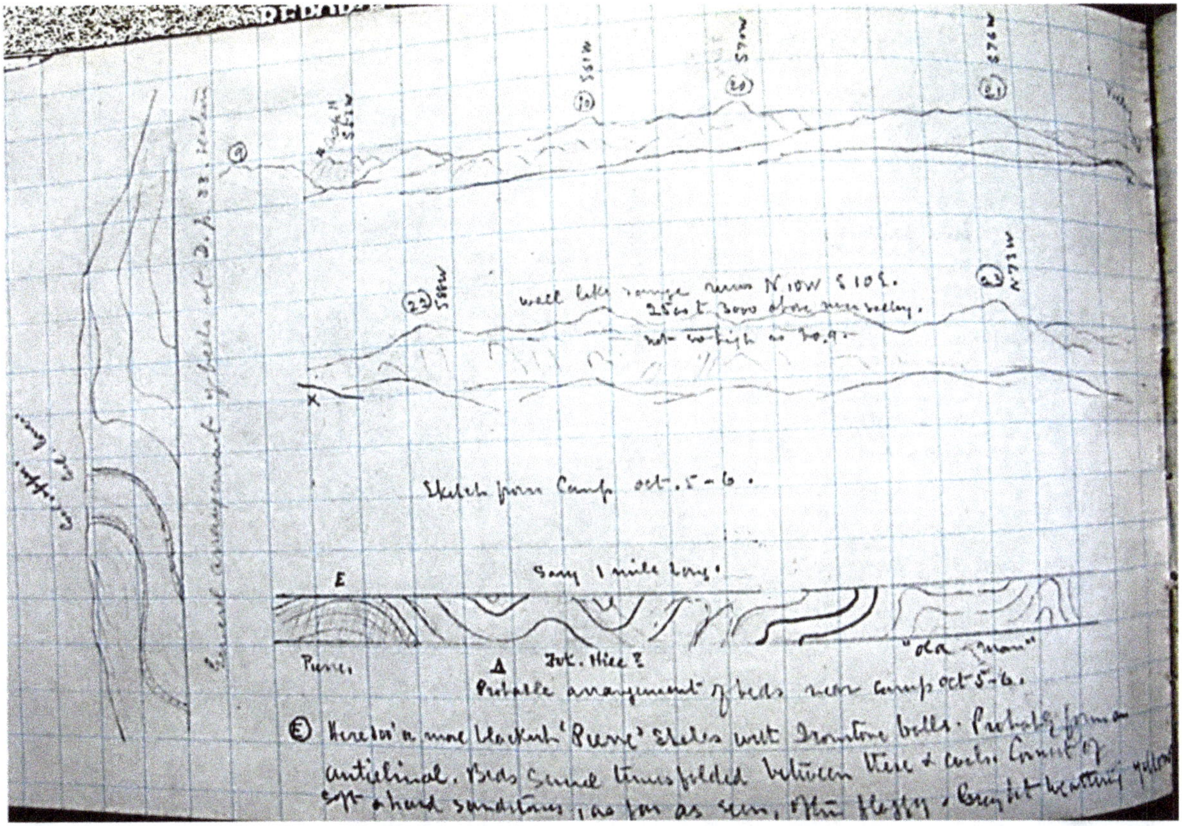

Dawson 1881 no 4 Geological Survey of Canada field notebooks.

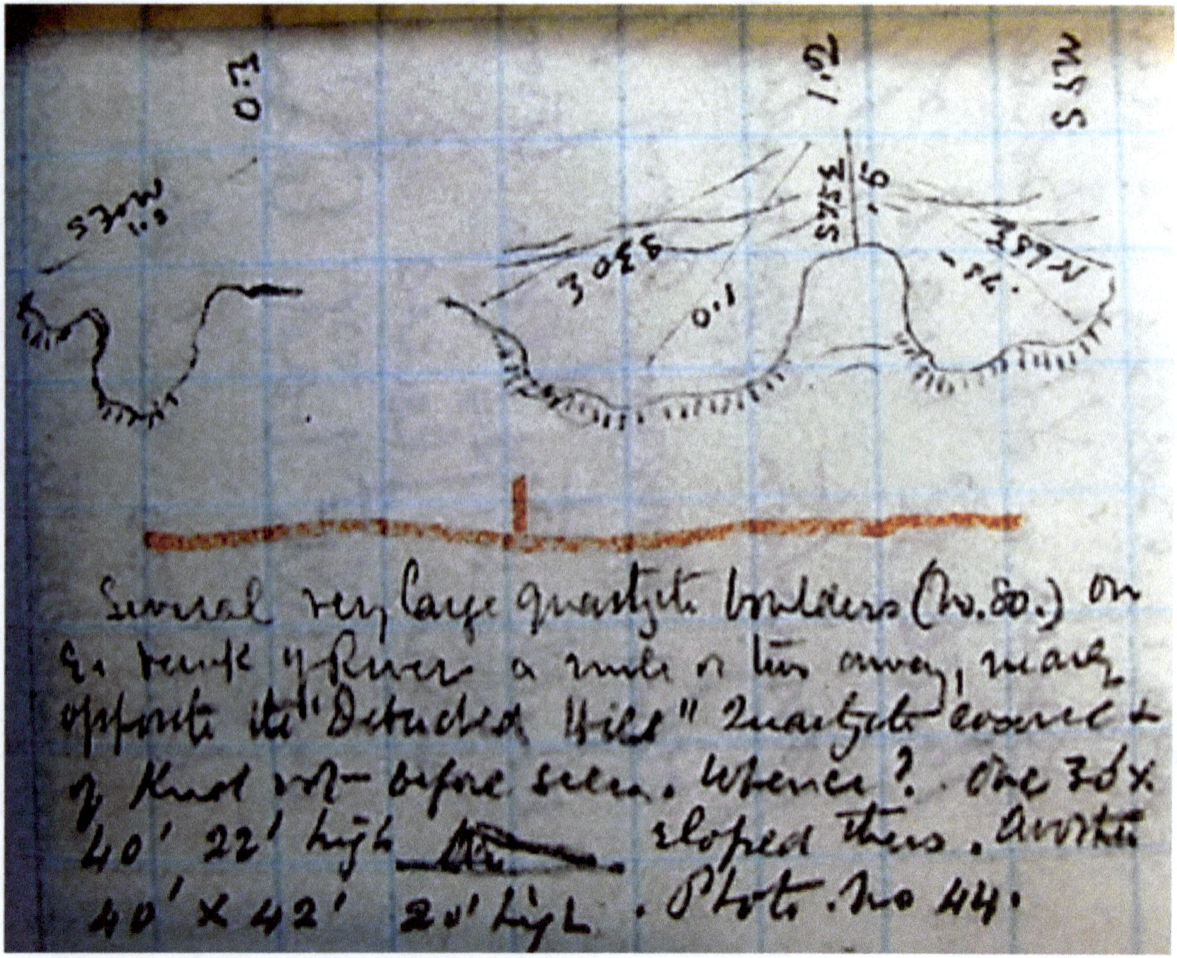

Dawson 1881. no. 2. 2746 Geological Survey of Canada field notebooks.

Dawson 1888 no2. Geological Survey of Canada field notebooks.

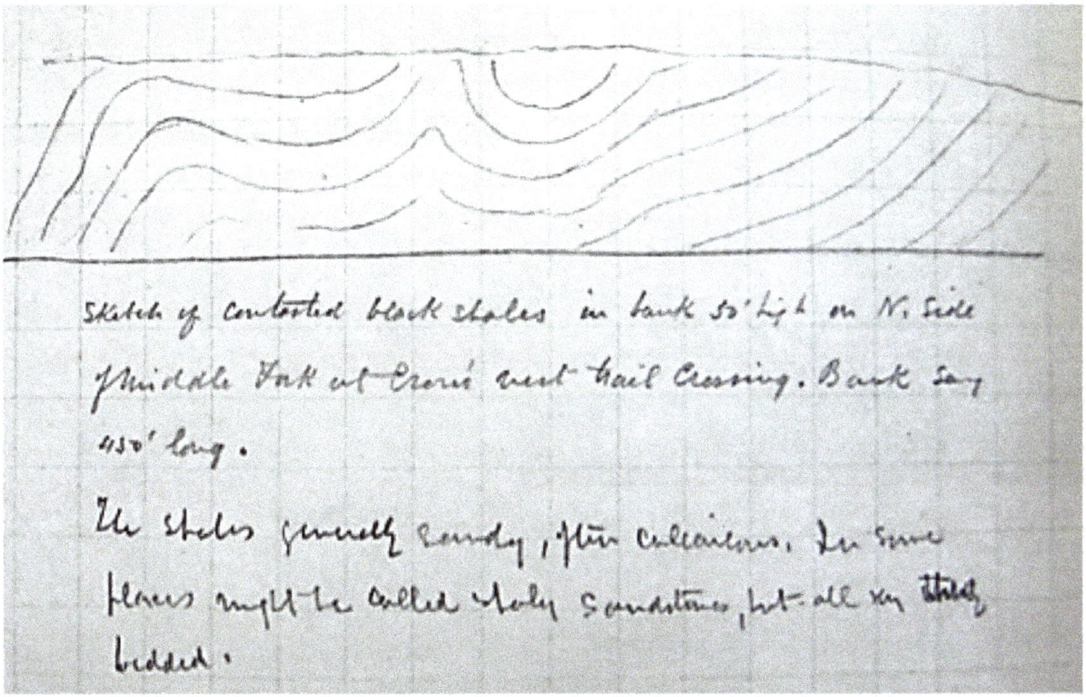

Dawson 1881 Geological Survey of Canada field notebooks.

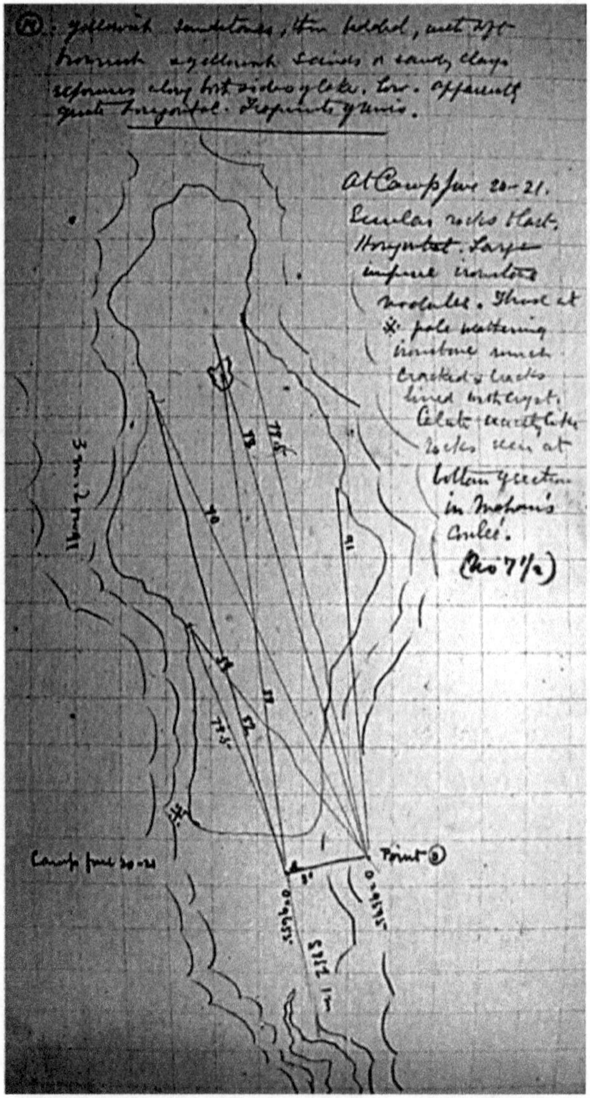

Dawson 1881 no5. Geological Survey of Canada field notebooks.

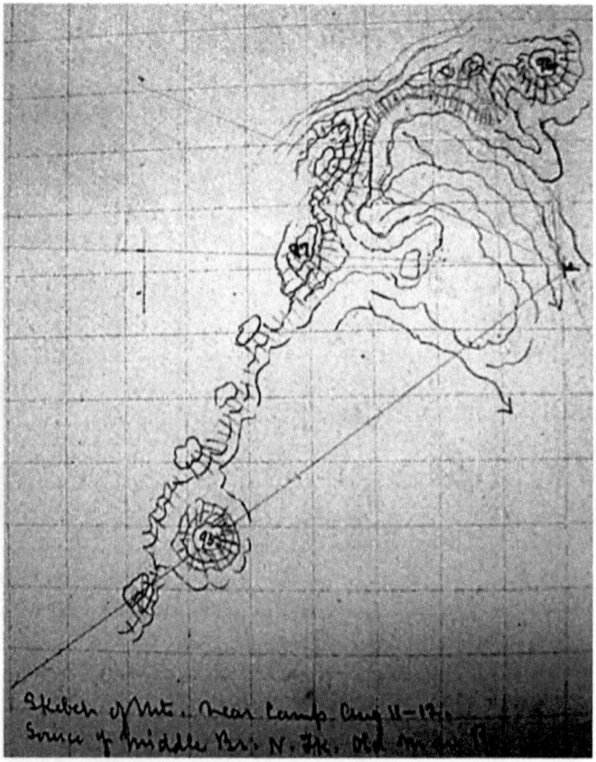

Dawson 1883 no 1. Geological Survey of Canada field notebooks.

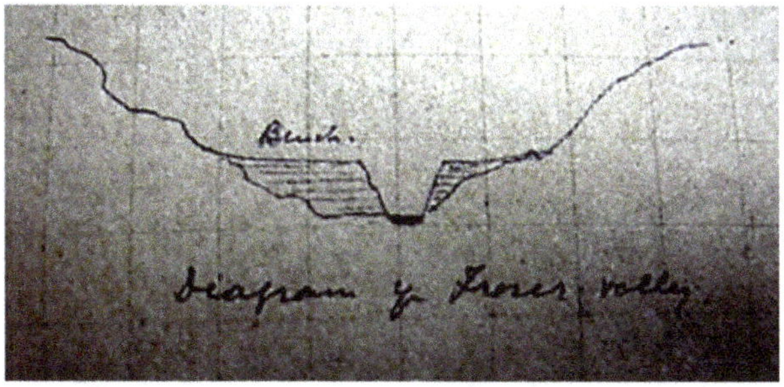

Dawson 1884 Geological Survey of Canada field notebooks.

Dawson 1890 no 3 Geological Survey of Canada field notebooks.

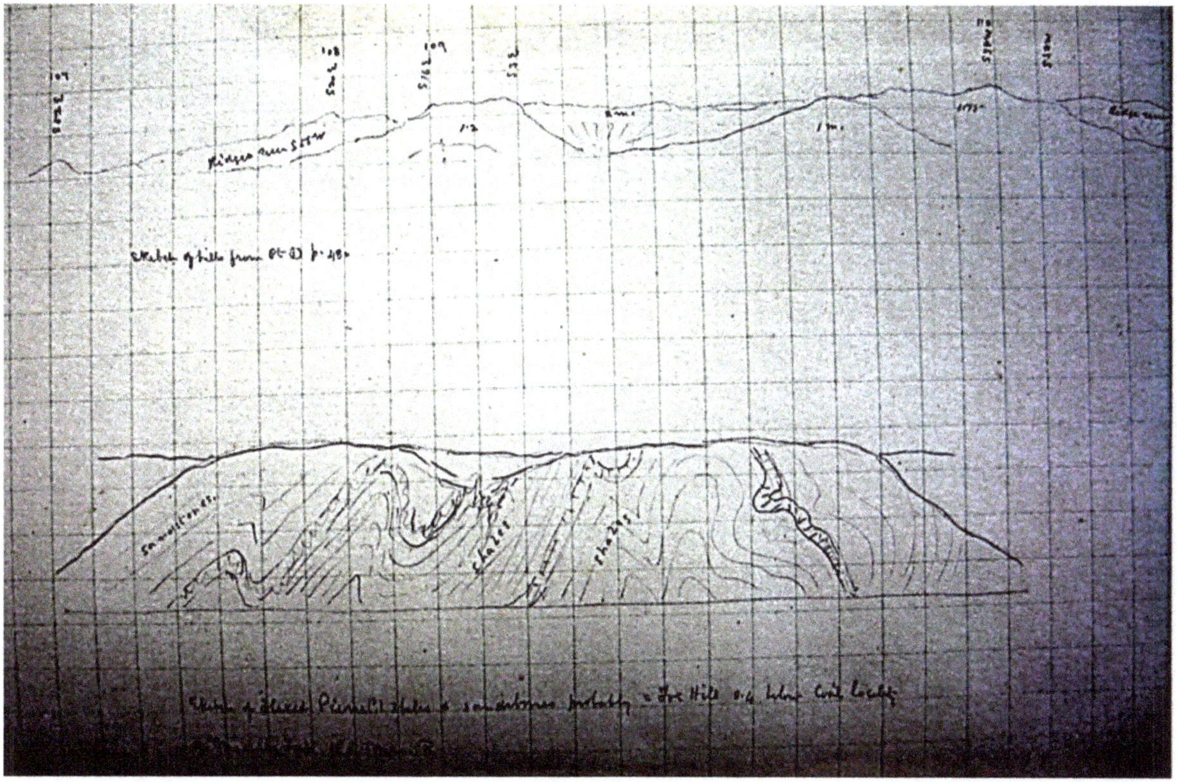

Dawson 1883 no 1 Geological Survey of Canada field notebooks.

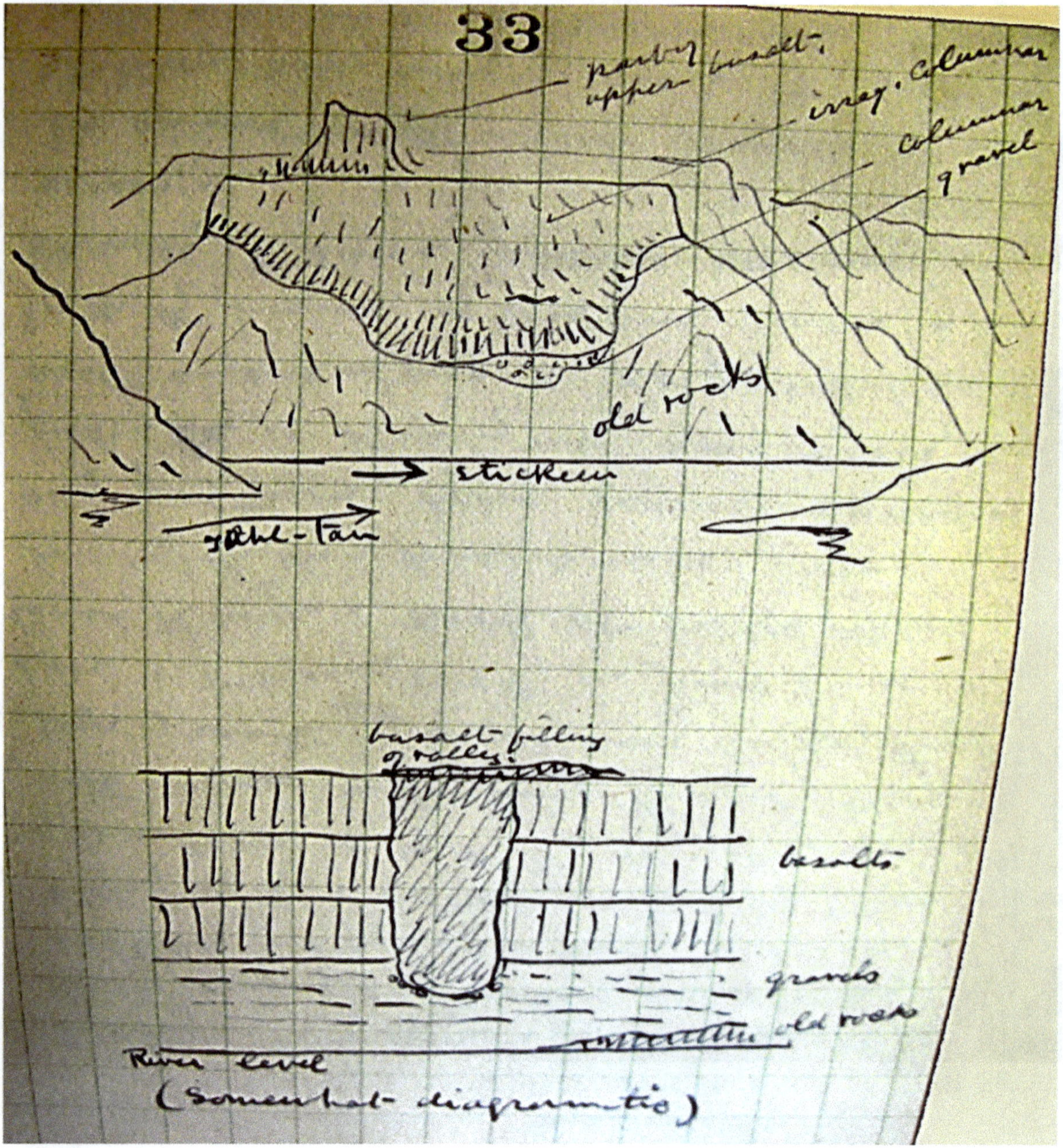

Dawson 1884 no 6 Geological Survey of Canada field notebooks.

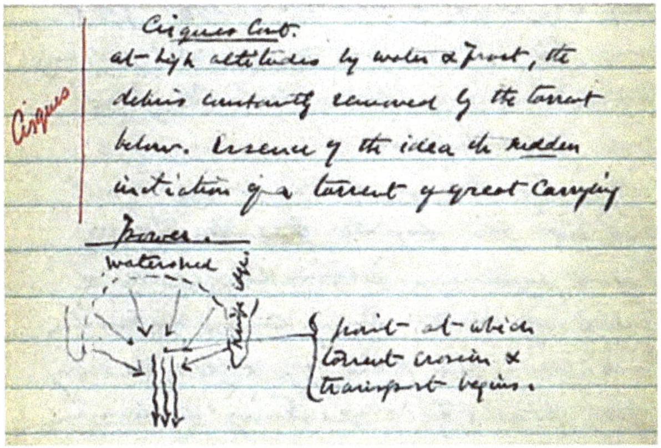

Dawson 1889. McGill niversity 1022 6 146 72

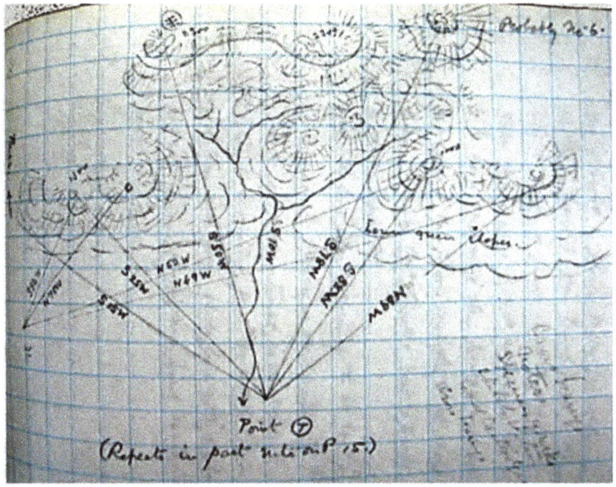

Dawson 1881. no. 2. 2746 Geological Survey of Canada field notebooks.

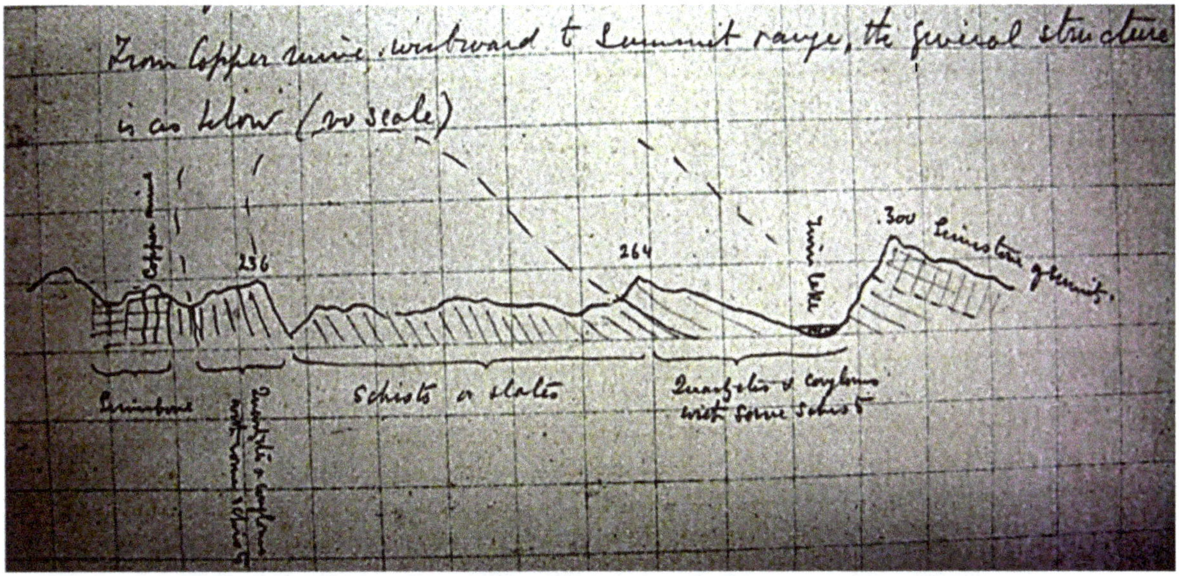

Dawson 1884 Geological Survey of Canada field notebooks.

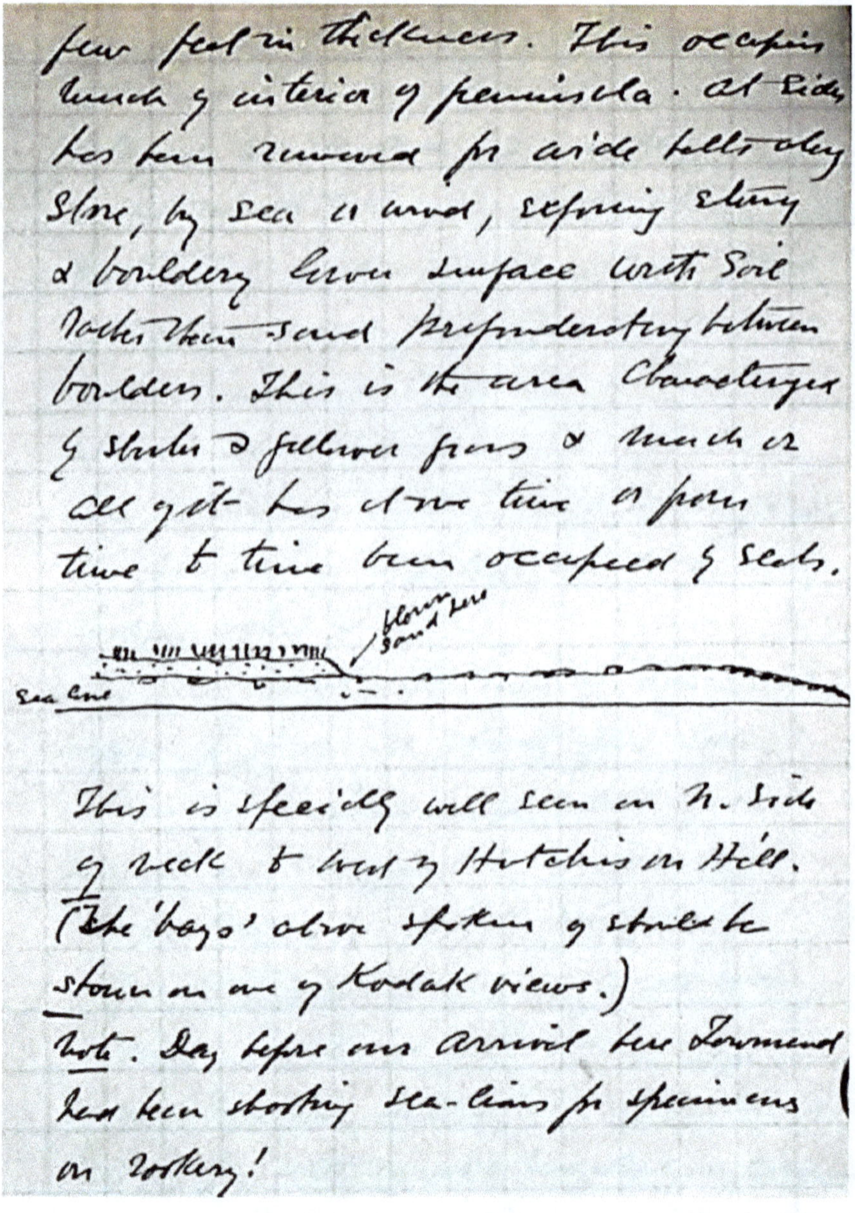

few feet in thickness. This occupies much of interior of peninsula. At sides has been removed fr aside tolerably Stone, by sea or wind, refusing stony & bouldery Grave surface with Soil rather than sand preponderating between boulders. This is the area characteristic of stoles & fallower grass & much or all of it has at one time or from time to time been occupied of seals.

This is specially well seen on N. side of neck to west of Hutchinson Hill. (The 'bays' above spoken of should be shown on one of Kodak views.)

Note. Day before our arrival here Townsend had been shooting sea-lions for specimens on rookery!

Dawson 1891. McGill niversity 1022 6 63 75 Sealing

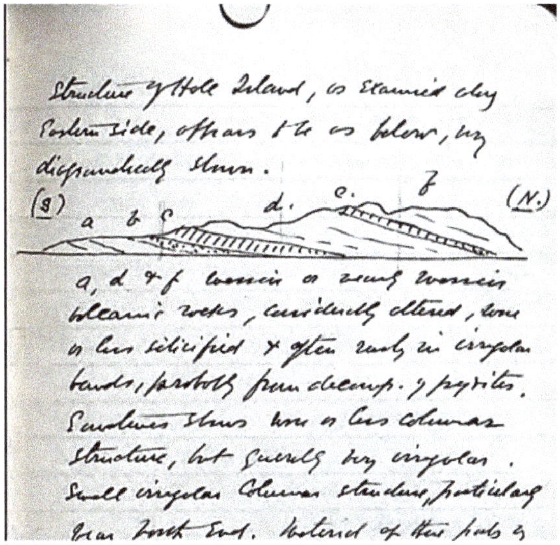
Structure of Hole Island, as examined along Eastern side, appears to be as below, very diagrammatically shown.

a, d & f breccia or scoriaceous breccia volcanic rocks, considerably altered, more or less silicified & often rusty in irregular bands, probably from decomp. of pyrites. Sometimes shows more or less columnar structure, but generally very irregular. Small irregular columnar structure, particularly near North End. Material of these parts of

Dawson 1891. McGill niversity

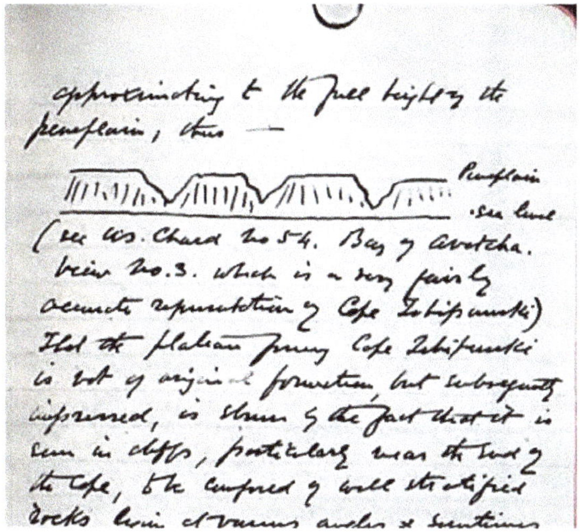
approximating to the full height of the peneplain, thus —

(see Ws. Chart No 54. Bay of Avatcha. View No.3. which is a very fairly accurate representation of Cape Tshipunski) That the plateau forming Cape Tshipunski is not of original formation but subsequently impressed, is shown by the fact that it is seen in cliffs, particularly near the end of the Cape, to be composed of well stratified rocks lying at various angles & sometimes

Dawson 1891. McGill niversity
1022 6 150 27

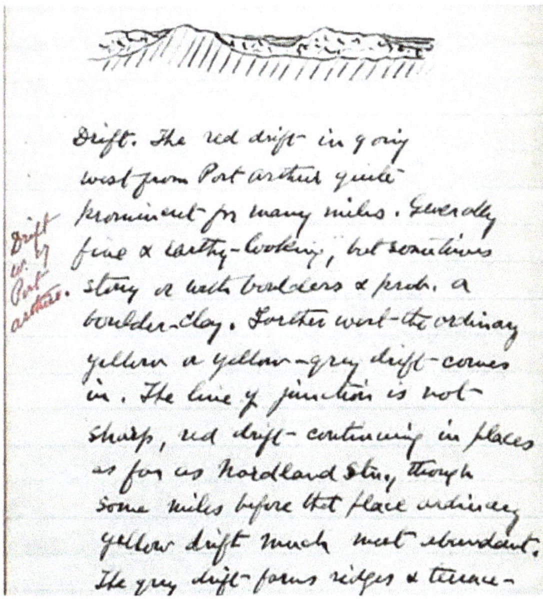
Drift. The red drift in going west from Port Arthur quite prominent for many miles. Generally fine & earthy-looking, but sometimes stony or with boulders & prob. a boulder-clay. Further west the ordinary yellow or yellow-grey drift comes in. The line of junction is not sharp, red drift continuing in places as far as Nordland Sta., though some miles before that place ordinary yellow drift much most abundant. The grey drift forms ridges & terrace-

Drift W. of Port Arthur.

Dawson 1890. McGill niversity
1022 6 147 6

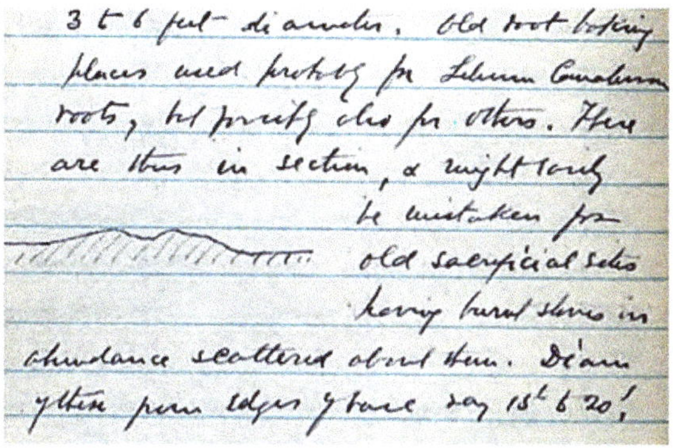
3 to 6 feet diameter. Old root looking places used probably for Lilium Canadense roots, but possibly also for others. These are thus in section, & might easily be mistaken for old sacrificial sites having burnt stones in abundance scattered about them. Diam often from edges of trace say 15' to 20'.

Dawson 1891. McGill niversity
1022 6 64 91

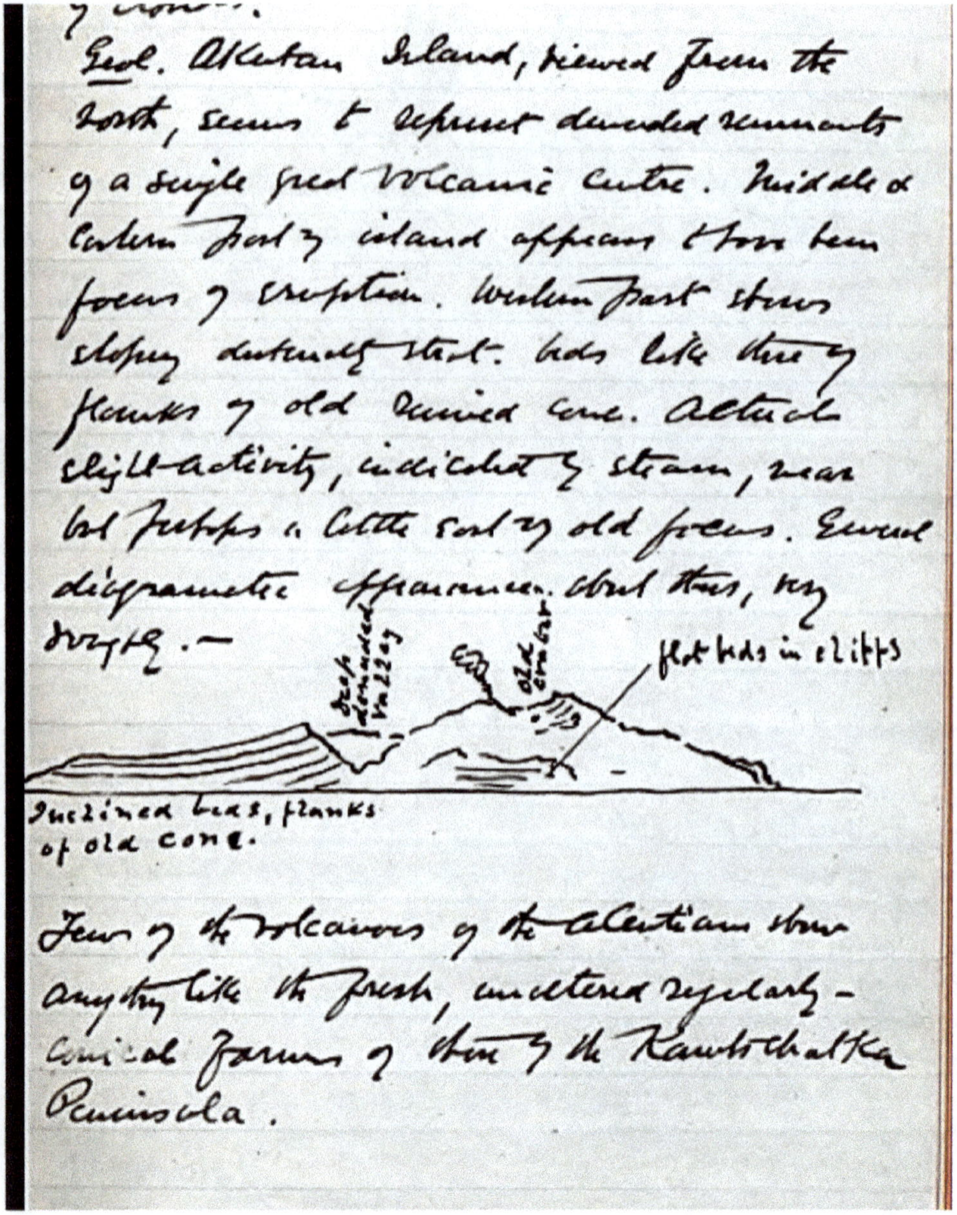

Geol. Akutan Island, viewed from the North, seems to represent denuded remnants of a single great volcanic centre. Middle & eastern part of island appears to have been focus of eruption. Western part shows sloping distinctly strat. beds like those of flanks of old denuded cone. Actual slight activity, indicated by steam, near but perhaps a little south of old focus. General diagrammatic appearance about thus, very roughly. —

Few of the volcanoes of the Aleutians show anything like the fresh, unaltered regularly-conical forms of those of the Kamtschatka Peninsula.

Dawson 1891. McGill niversity 1022 6 150 55

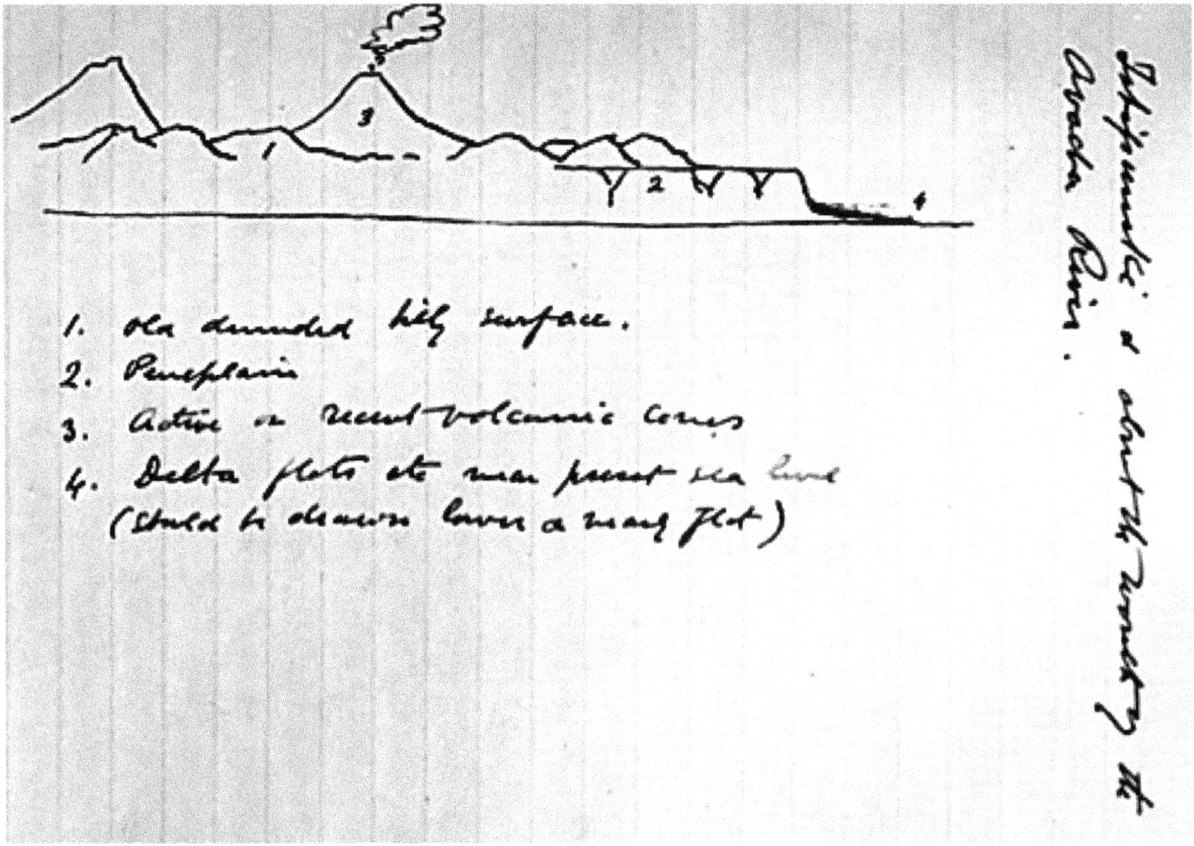

Dawson 1891. McGill niversity 1022 6 150 29

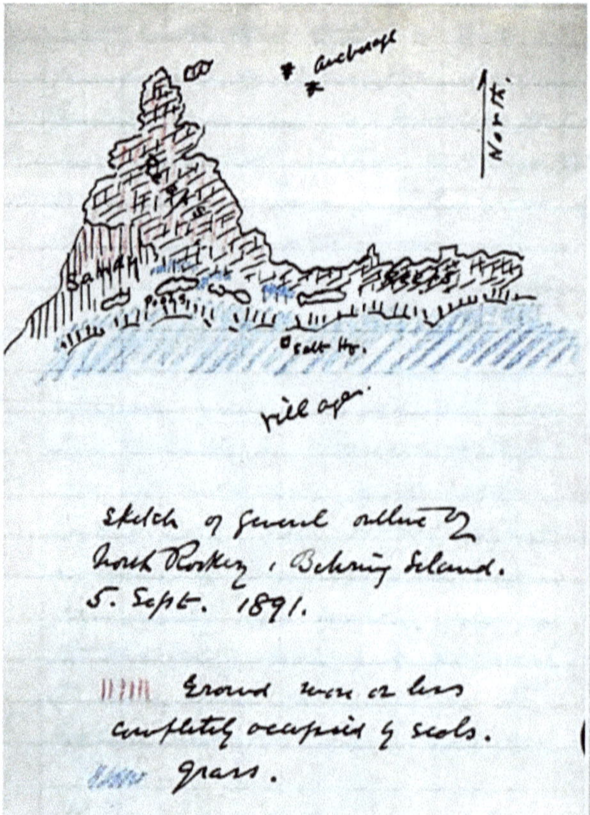

Dawson 1891. McGill niversity 1022 6 63 L II 131 riginal is coloured. Sealing

Dawson 1894 no 6. Geological Survey of Canada field notebooks.

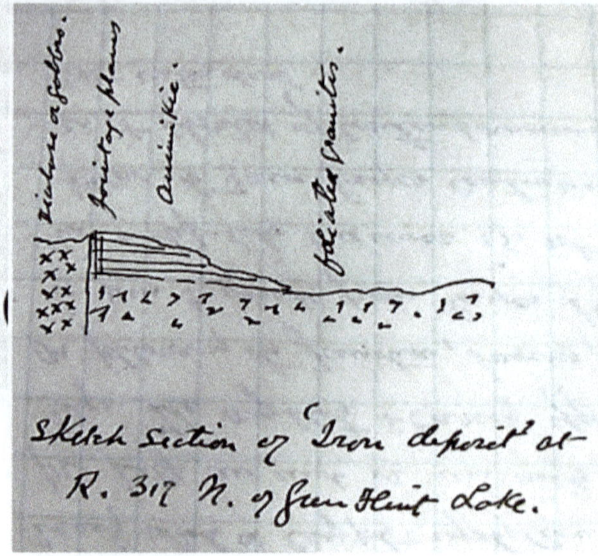

Dawson 1894 no 6. Geological Survey of Canada field notebooks.

the westward of high Point of Cape Upright, rocky cliffs forming part between this beach & anchorage, appear darker in colour, possibly basaltic. At the extreme of the Point a bed or layer of some thickness of greyish yellow colour & very probably of sedimentary rock such as sandstone or shale intercalated in the darker rock & dipping southward ∠ 15° about. Under surface seems to rest not quite conformably on layers of lower dark rock, & upper surface irregular as though ploughed up by upper dark rock, somewhat as below.—

Dawson 1891. McGill niversity 1022 6 64 59

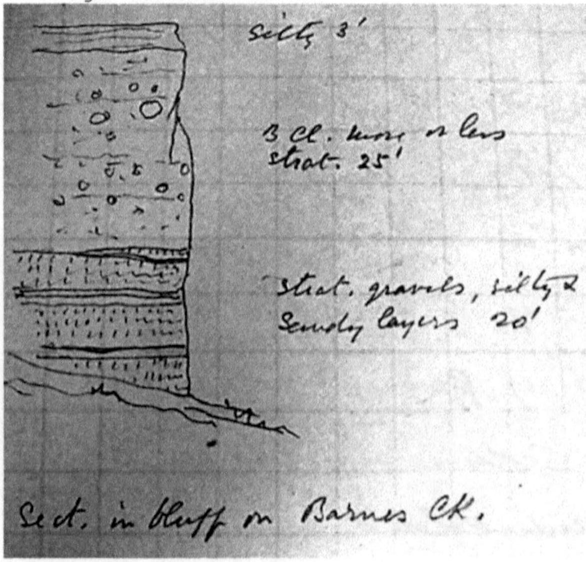

Dawson 1894 no 1 Geological Survey of Canada field notebooks.

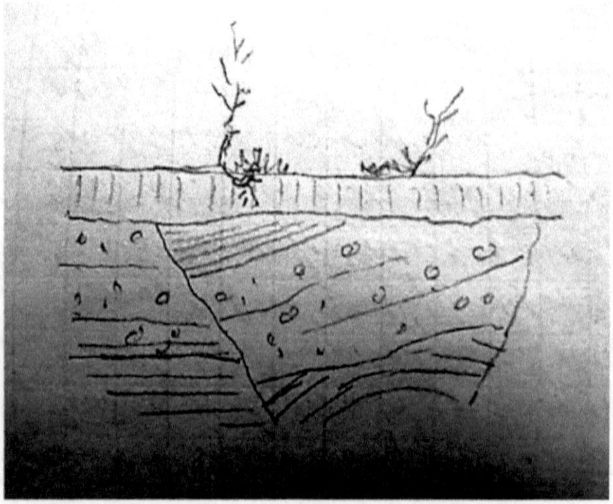

Dawson 1894 no 5 Geological Survey of Canada field notebooks.

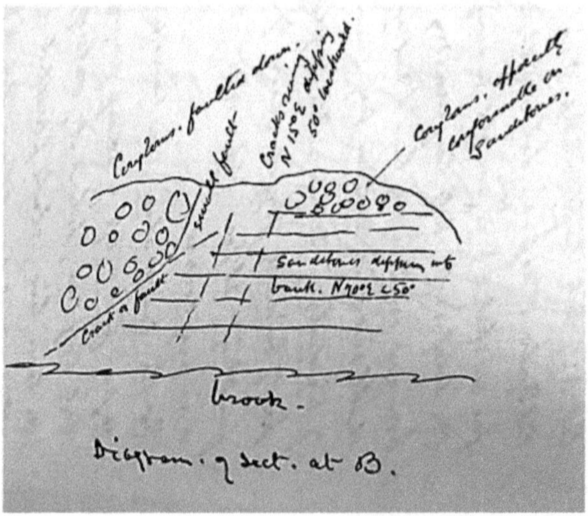

Dawson 1894 no 5 Geological Survey of Canada field notebooks.

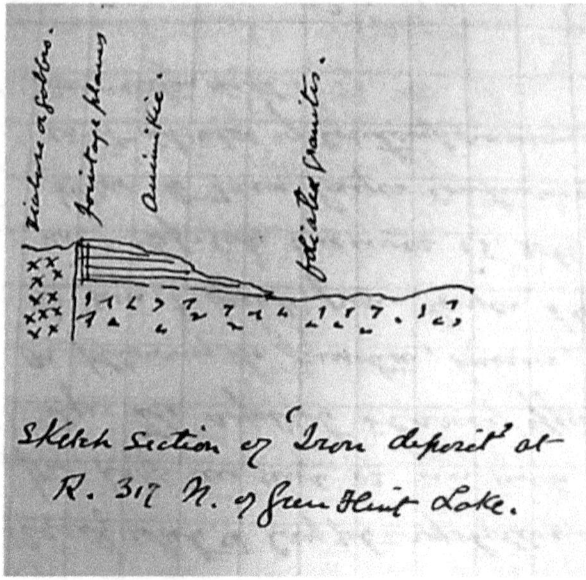

Dawson 1894 no 6 Geological Survey of Canada field notebooks.

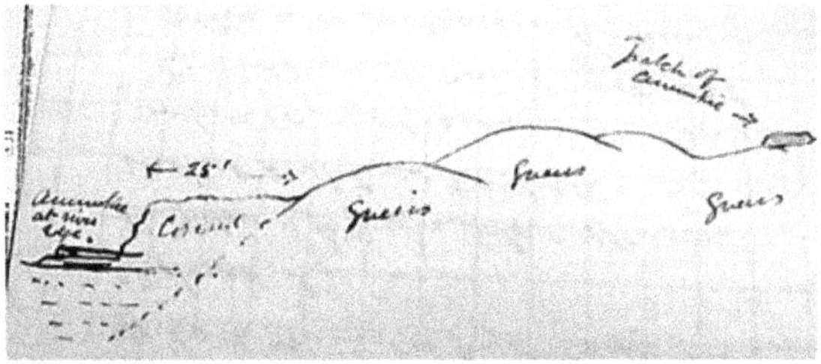

Dawson 1894 no 6. Geological Survey of Canada field notebooks.

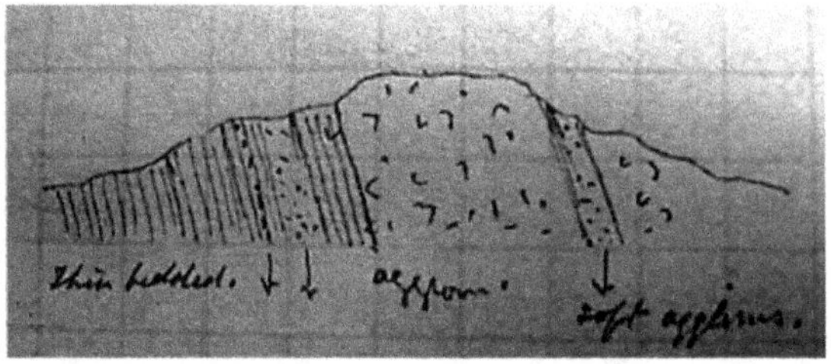

Dawson 1894 no 1 Geological Survey of Canada field notebooks.

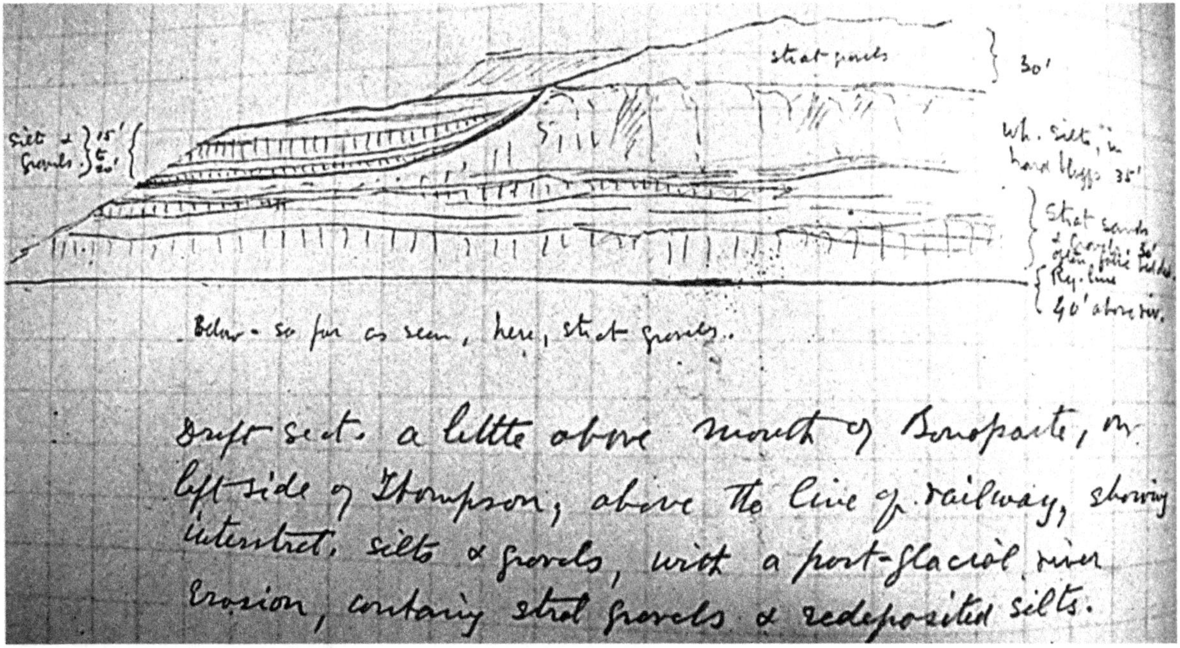

Dawson 1894 no 1 Geological Survey of Canada field notebooks.

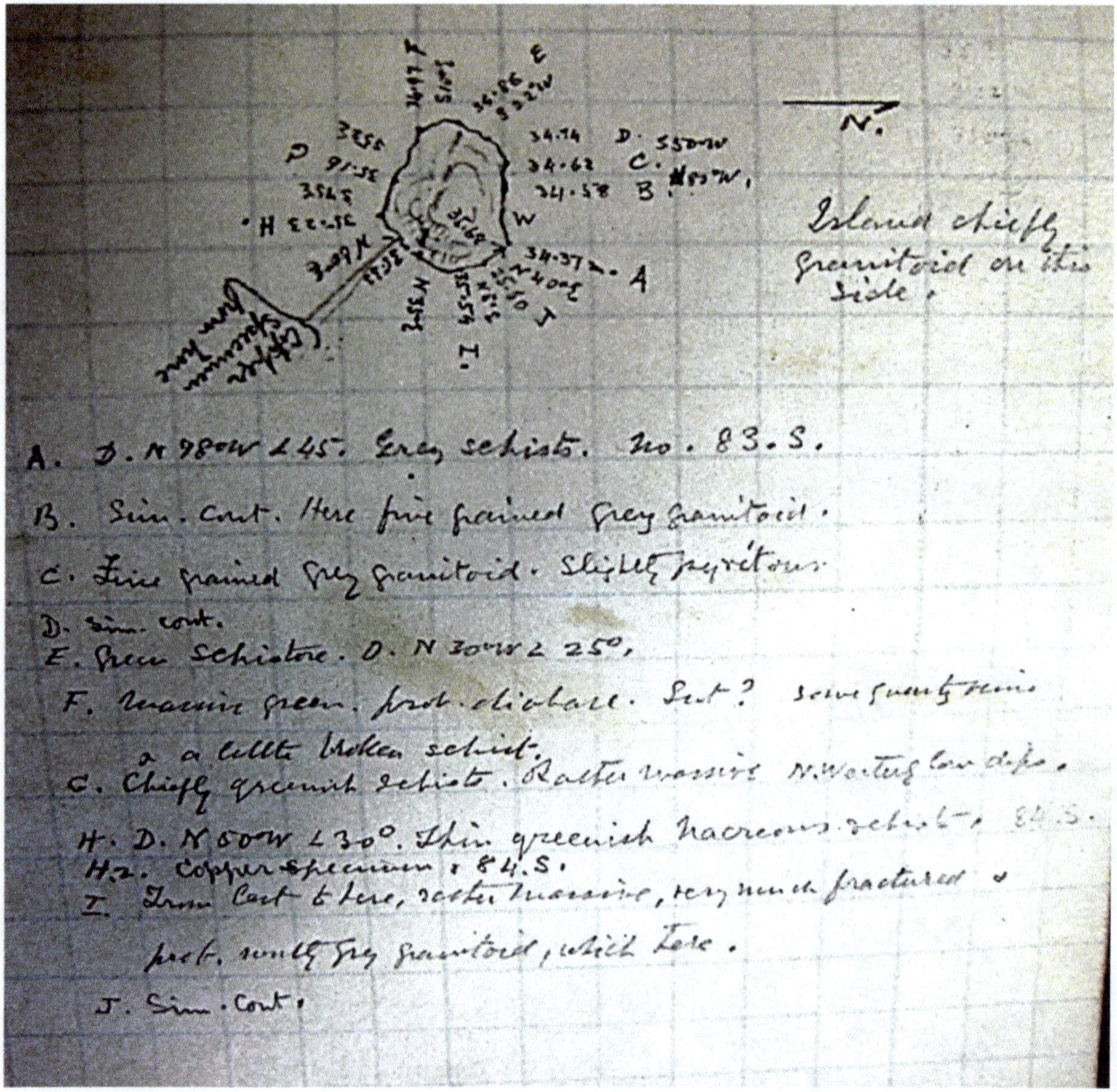

N.

Island chiefly granitoid on this side.

A. D. N 78°W ∠45. Grey schists. No. 83.S.

B. Sim. cont. Here fine grained grey granitoid.

C. Fine grained grey granitoid. Slightly pyritous.

D. Sim. cont.

E. Green schistose. D. N 30°W ∠ 25°.

F. Massive green. prob. diabase. Sut? some quartz veins a little broken schist.

G. Chiefly greenish schist. Rather massive N. Westerly low dips.

H. D. N 60°W ∠30°. Thin greenish nacreous schist. 84.S.

H.2. Copper specimen. 84.S.

I. From cont to here, rather massive, very much fractured & prob. mostly grey granitoid, which here.

J. Sim. Cont.

Dawson 1894 no 4 Geological Survey of Canada field notebooks.

Dawson 1898. McGill niversity 1022 6 155 84

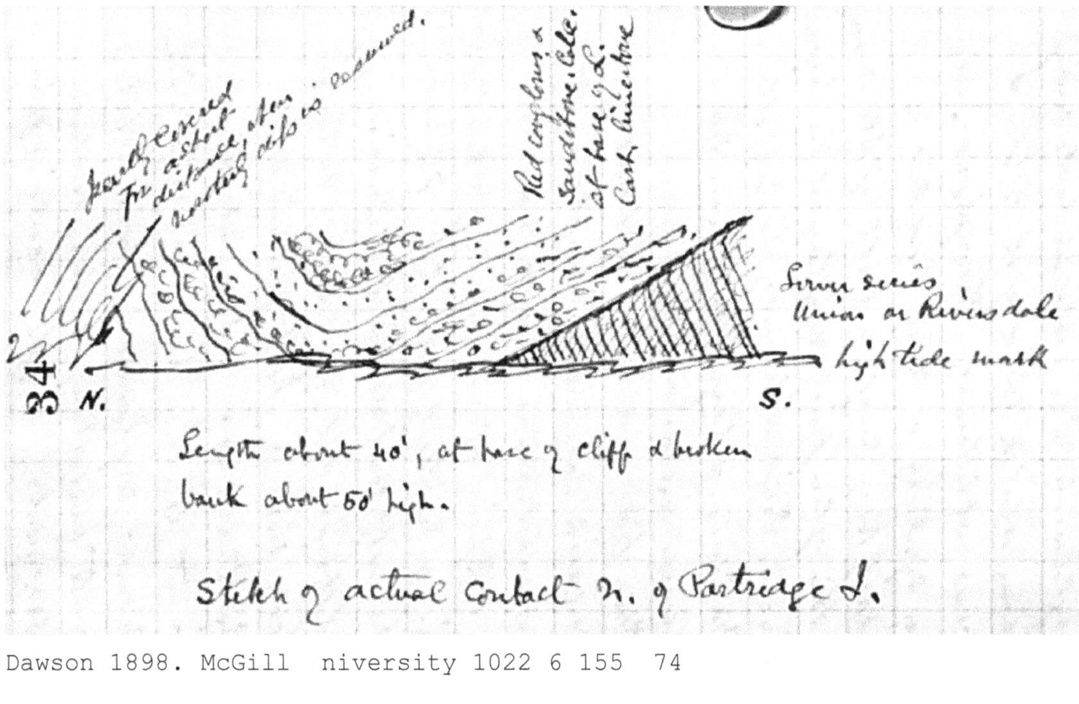

Dawson 1898. McGill niversity 1022 6 155 74

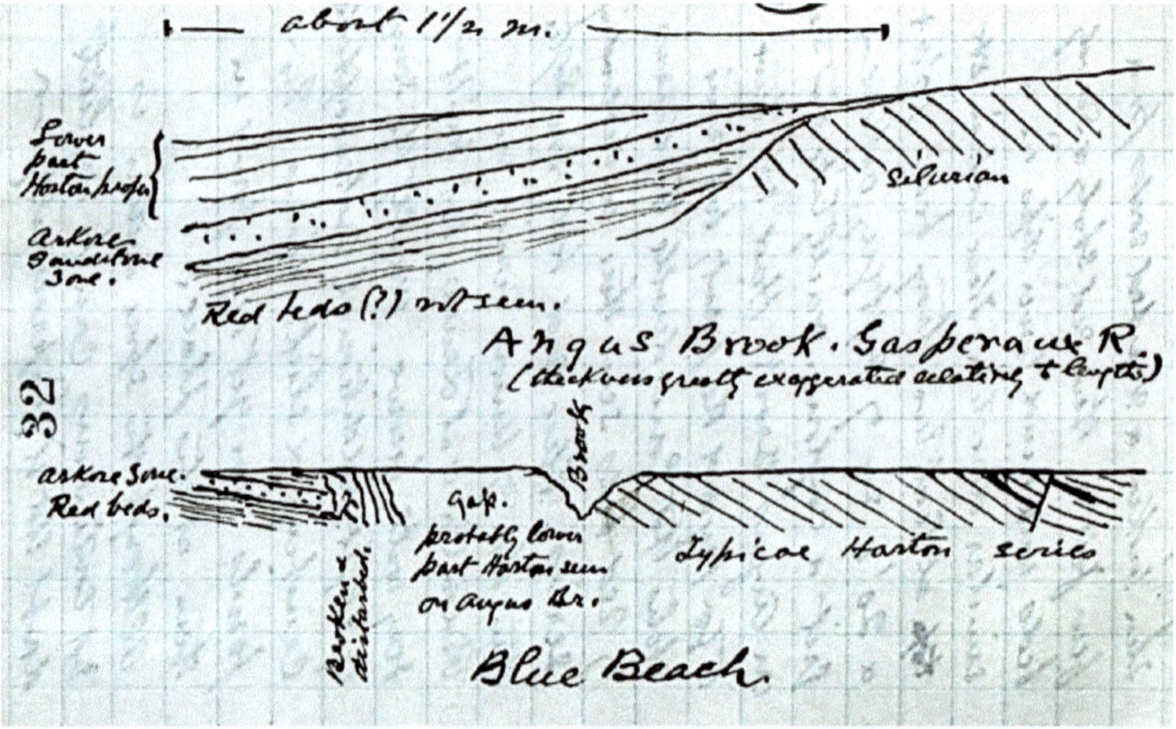

Dawson 1898. McGill niversity 1022 6 155 70

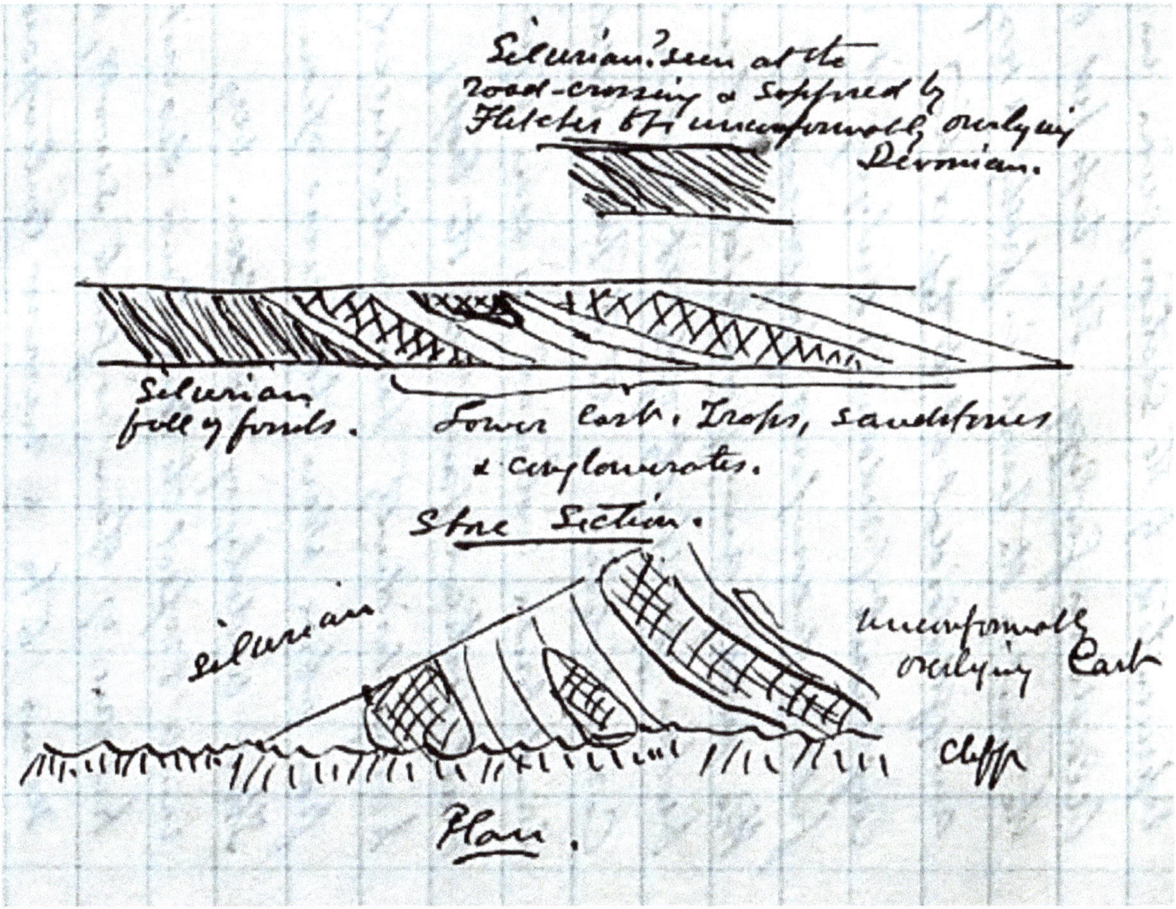

Dawson 1898. McGill niversity 1022 6 155 92

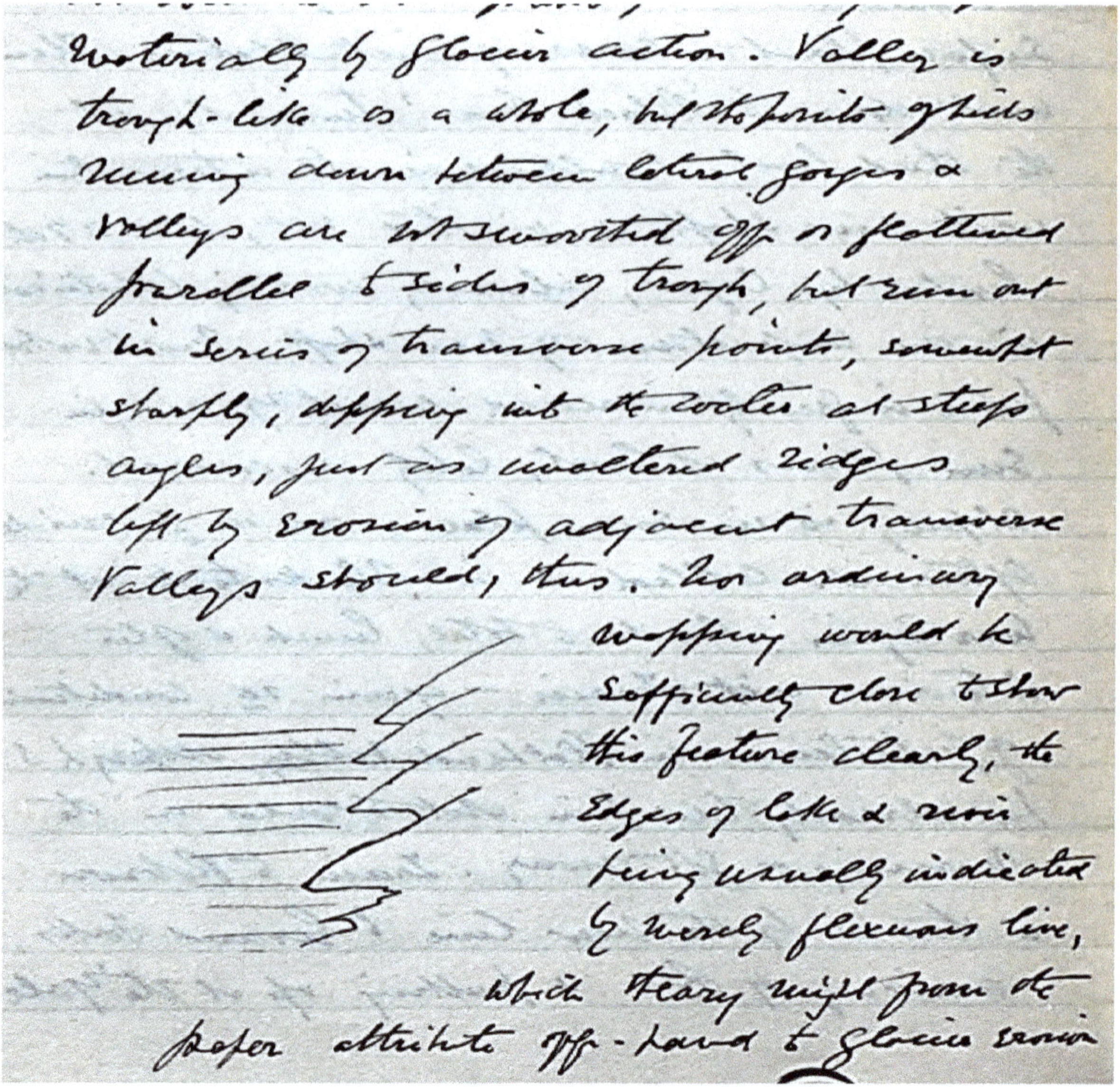

materially by glacier action. Valley is
trough-like as a whole, but the points of hills
running down between lateral gorges &
valleys are not smoothed off or flattened
parallel to sides of trough, but run out
in series of transverse points, somewhat
sharply, dipping into the valley at steep
angles, just as unaltered ridges
left by erosion of adjacent transverse
valleys should, thus. No ordinary
mapping would be
sufficiently close to show
this feature clearly, the
edges of lake & river
being usually indicated
by merely flexuous line,
which theory might from the
paper attribute off-hand to glacier erosion

Dawson 1899. McGill niversity 1022 6 157 90

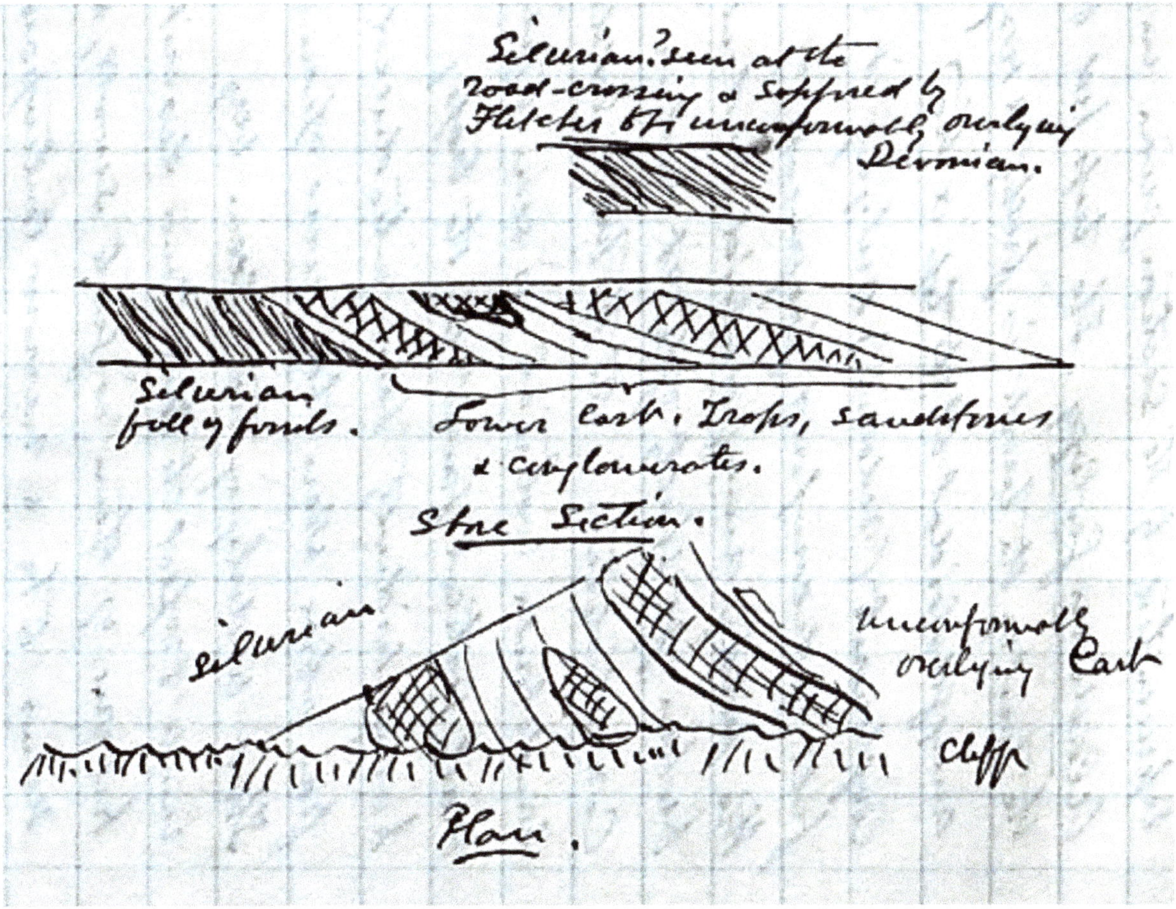

Dawson 1898. McGill niversity 1022 6 155 92

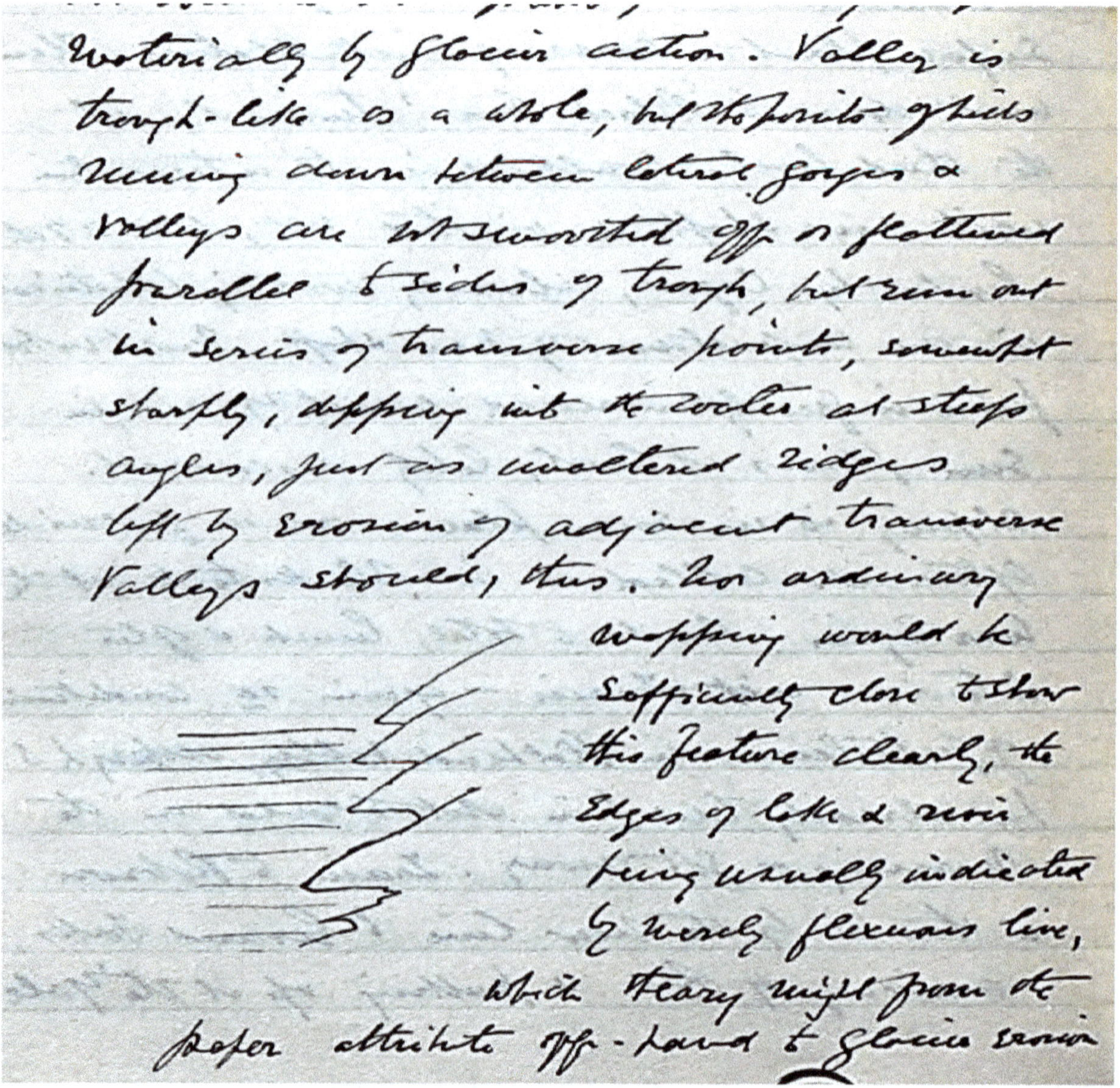
materially by glacier action. Valley is trough-like as a whole, but the points of hills running down between lateral gorges & valleys are not smoothed off or flattened parallel to sides of trough, but run out in series of transverse points, somewhat sharply, dipping into the water at steep angles, just as unaltered ridges left by erosion of adjacent transverse valleys should, thus. No ordinary mapping would be sufficiently close to show this feature clearly, the edges of lake & river being usually indicated by merely flexuous line, which theory might from the paper attribute off-hand to glacier erosion

Dawson 1899. McGill niversity 1022 6 157 90

Miscellaneous

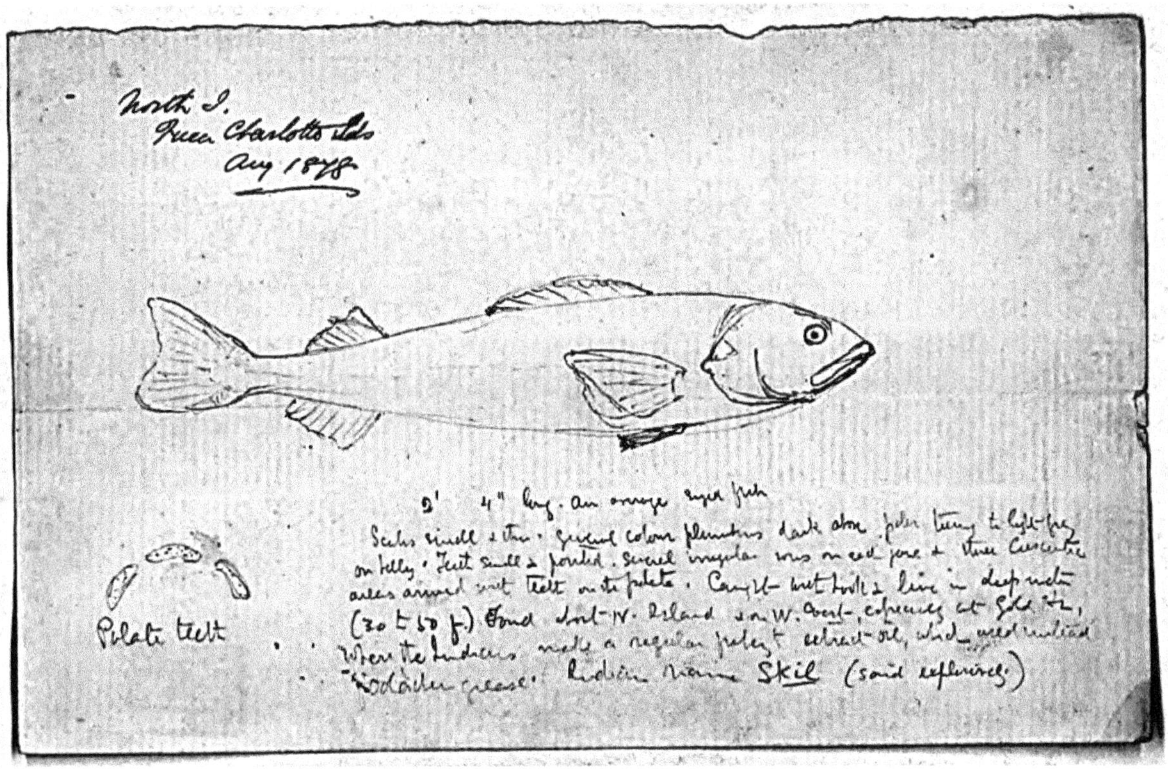

Dawson 1878. McGill niversity 1022 6 51 15

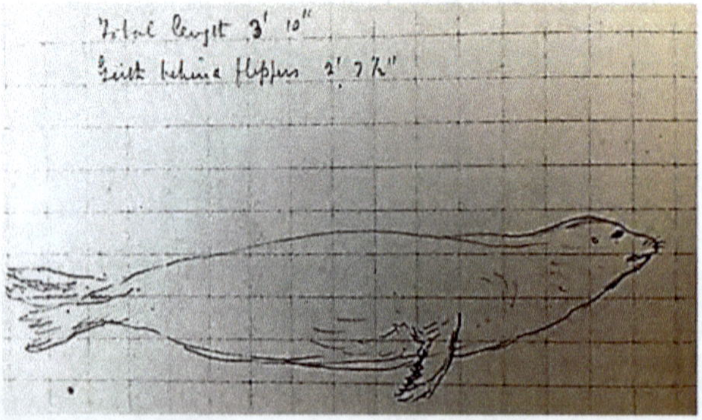

Dawson 1885. n3. ational Library
Geological Survey of Canada field
notebooks.

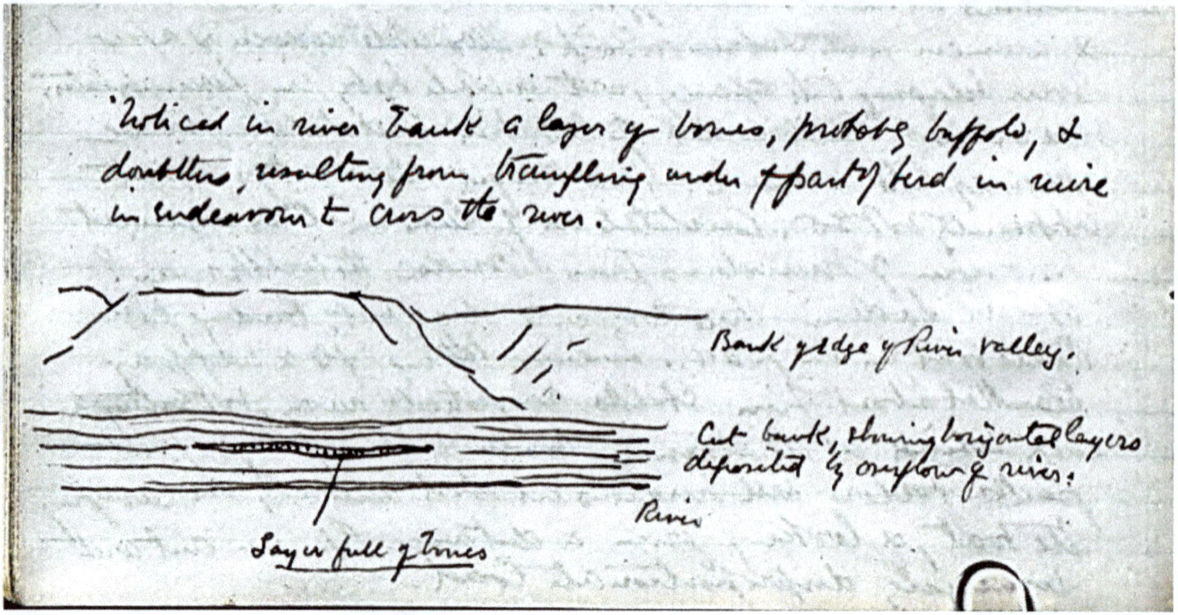

Dawson 1881. McGill niversity 1022 6 137 32

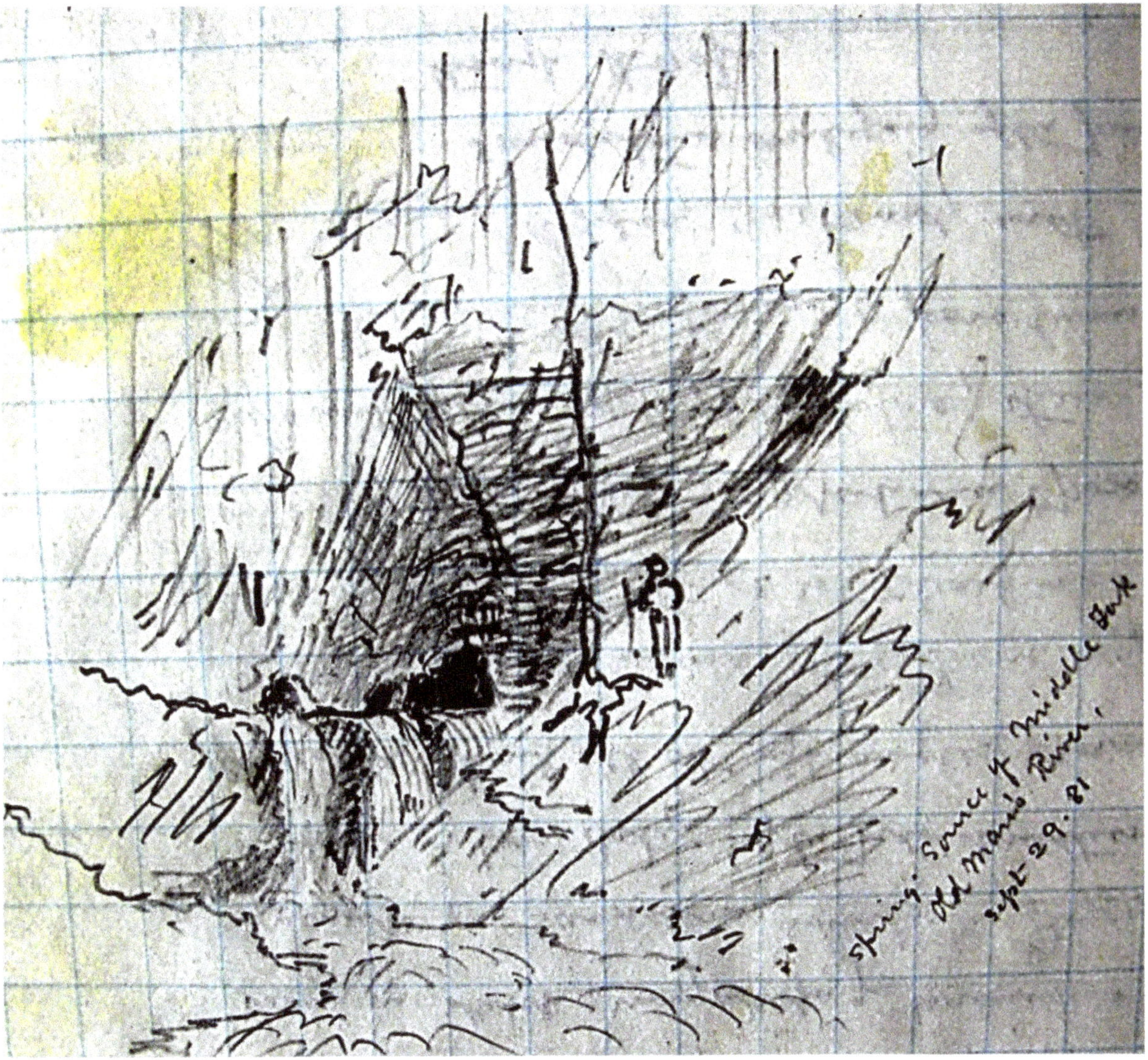

Dawson 1881 no 4. Geological Survey of Canada field notebooks.

Dawson 1881. McGill niversity 1022 6 137 150

www.ingramcontent.com/pod-product-compliance
Lightning Source LLC
Chambersburg PA
CBHW061218240926
51846CB00006B/1267
* 9 7 8 1 9 9 8 3 2 1 4 5 2 *